BUILDING
Doors & Gates

BUILDING
Doors & Gates

Instructions, Techniques and Over 100 Designs

**ALAN & GILL
BRIDGEWATER**

STACKPOLE
BOOKS

To John and Amy.
We look forward to a great many more walks, talks, E-mails, and days out—all those doors we have opened, gates we have stumbled through, and stiles we have clambered over.

Copyright © 1999 by Stackpole Books

All rights reserved, including the right to reproduce this book or portions thereof in any form or by any means, electronic or mechanical, including photocopying, recording, or by any information storage and retrieval system, without permission in writing from the publisher. All inquiries should be addressed to Stackpole Books, 5067 Ritter Road, Mechanicsburg, Pennsylvania 17055.

FIRST EDITION

Cover design by Wendy Reynolds
Cover photograph by Ian Parsons
Interior line drawings by the authors

Library of Congress Cataloging-in-Publication Data

Bridgewater, Alan.
　　Building doors and gates / Alan and Gill Bridgewater.—1st ed.
　　　　p.　　cm.
　　ISBN 0-8117-2678-9
　　1. Doors—Design and construction.　2. Gates—Design and construction.
　I. Bridgewater, Gill.　II. Title.
TH2278.B75　　　1999
694'.6—dc21　　　　　　　　　　　　　　　　　　　98-27518
　　　　　　　　　　　　　　　　　　　　　　　　　　　CIP

ISBN 978-0-8117-2678-8

Contents

Introduction	vii
Part 1: DOORS	1
Anatomy of the Door and Door Joints	5
Generic Door Types	24
Making a Ledged Door	40
Making a Four-Panel Door	45
Hanging and Fitting a Door	54
Door Designs	72
Part 2: GATES	103
Anatomy of the Gate and Gate Joints	107
Generic Gate Types	117
Making a Five-Bar Sussex Field Gate	133
Making a Framed Oak Garden Gate	140
Hanging a Field Gate	145
Gate Designs	150
Metric Conversions	165
Index	166

Introduction

Doors and gates figure highly in most cultures. They are much more than simply a means of sealing an opening—the front door to a house, the main door to a church, the gate to a garden, the gate at a frontier, the gate at the front entrance to a multinational corporation, or whatever—they are symbols. Doors and gates have come to symbolize power, permanence, authority, and property. If you see a large, well-made oak door or paneled door, your mind immediately begins to play word-association games—door, strong, stable, family, home, security. It's not by accident that the U.S. president is filmed in front of the door to the White House, and the prime minister of Britain is photographed in front of the door to Number 10. The media are forever using shots of doors and gates as stand-ins for corporation, industries, and powers. When they can't get to see the main players, they settle for filming the entrances. And doors and gates are always being used figuratively—doors are slammed in faces, doors are opened, talks are held behind closed doors, someone gets his foot in the door, and of course, we all know that it's useless to shut the stable door after the horse has bolted.

Doors and gates are much more than physical barriers; they are powerful statements, and as such, architects put a huge amount of effort into designing them. They are the first things that strike visitors' eyes when they see a building. In many ways, the door or gate tells you all you need to know about those living and working within. Just think back to your first day at school or your first job interview or the first meeting with your partner at his or her home. One of the first things that you did before entering the building—perhaps unknowingly—was to study the gate, door, or entranceway to see if it could tell you something about the people within. Was the door welcoming or intimidating? Was it painted a dark color? Or was it simply huge?

Doors and gates have evolved over many hundreds of years, and they are unique examples of the joiner's art. When you think that doors and gates are forever being opened and closed, exposed to all weathers, and generally being kicked, slammed, hammered on, and all the rest, it's plain to see that, perhaps more than any other piece of woodwork, they need to be designed and built with an extraordinary amount of care. If you look closely at a traditional door or gate, you will see that it is an amazingly complex structure, with every joint, detail, and component having evolved to fulfill a specific function. The stiles need to be heavy enough to take the hinges and the catches, the panels needs to be strong and stable yet lightweight, the mortise and tenon joints must be different from one end of the door to the other, exterior doors have to have drips and bars to keep out the rain, field gates have to swing freely on their hinges, and so forth. Doors and gates must be designed so that they can expand and contract with the ever-changing humidity without splitting, warping, or in any way becoming overstressed. On top of all this, the door or gate has to send out the right message. It's no good building a massive oak structure all bound and studded with iron if all you need is a door for the toilet. And what does a badly painted, warped front door tell visitors?

When we first had the notion to write this book, we asked the question why, in this day and age, when prehung flush doors can be purchased off the shelf and fitted in minutes and ready-made gates are just as available, would anyone want to build a door or gate. Well, of course, the answer is wonderfully simple: A traditional door is much more than a slab of wood to fill a hole, and a traditional five-bar field gate is more than just a collection of rails. They are truly desirable works of art. A traditional hand-built door is a uniquely beautiful item—it looks good, it is proven, and it will last a lifetime.

The more we researched the subject, the more we came to realize that a person's individuality can be expressed as much with a door or a gate as with clothes, a car, a painting, or whatever. We soon came to realize that hand-built doors and gates are special—more pieces of sculpture than items of functional woodwork. Build an oak door for the front entrance to your house, a gate for your drive, or a romantic bower gate at the end of your garden, and you immediately begin to unlock the potential of your home and your personality.

In this book, we tell you about all manner of traditional doors and gates. We describe the component parts, the joints and the structures, and how to design, make, and hang a selection of doors and gates. And hundreds of designs and details are provided to help you on your way. After you have read this book, you will probably never be able to look at a door or gate in the same way again.

So what are you going for? An oak door in the grand Gothic tradition? A kissing gate? A medieval battened and ledged door with handmade ironwork? Or . . .

Part 1

DOORS

Of all architectural woodwork, door making is one of the most taken for granted, one of the most necessary, and one of the most complicated forms of construction. We open doors, we slam them, we beat on them, and we generally use and abuse them to the extent that we forget they are there. The only time most of us give any thought to the doors in our homes is when they refuse to open or close.

At one level, a door is a door is a door—just a slab of wood that fills a hole—but that's only the start of it. While it is true to say that a door performs two simple functions—it seals off an opening, and it opens and closes to allow easy entry and exit—the secondary function of a door, some would say the most important function, is that it tells you something about the building.

The main entrance doors to our homes are traditionally built to make the biggest possible statement. Your front door might be painted a bright color to show that you are different from your neighbors, it might be made from exotic wood and covered in brass fittings to show that you have wealth, it might be covered in complicated locks to show that your house is especially secure, or it might be embellished with classical pillars to show that you have taste. And so it is with public buildings; the church, the synagogue, the bank, and the courthouse all have front doors that make a statement.

While a door undoubtedly needs to be attractive, it also needs to suit its purpose. Your front entrance door must be easy to open and close, nice and snug to keep out the weather, and strong enough to keep out intruders, and it must do all this year in and year out, without shrinking, twisting, cracking, or otherwise failing. The bringing together of all these needs and functions adds up to a uniquely exciting woodworking problem.

DESIGN AND CONSTRUCTION

Traditionally, doors are named in accordance with their form of construction, their position within the building, or both. A door might be described as being a ledged, ledged and braced, framed and paneled, or paneled main entrance door; a framed and glazed side door; a ledged shed door; an exterior door; a high-security door; and so forth.

When you are considering designing and making a door, before you ever put tools to wood, you need to ask yourself the following questions.

- Where do you want the door to be placed—is it a front door, a back door, or an interior door?
- Does the door have to be extra secure, with special fittings or special glass?
- Does the door have to conform to various health or safety requirements or be made of special fireproof materials? Must it have spring hinges that swing it back into the closed position?
- Does the door need to be a nonstandard size—for example, accessible to wheelchairs?
- If it is an exterior door, is it exposed to lots of

A framed and paneled door in the classical Western tradition with a carved surround in London, England. This beautiful exterior door has it all: solid molded framing and raised panels, a fanlight above, and a broken segmental pediment with finials and swags supported on Greek Corinthian fluted pilasters. This is the type of door that set the scene for all that was to follow. If you are looking for the door to beat all doors, perhaps this is the one.

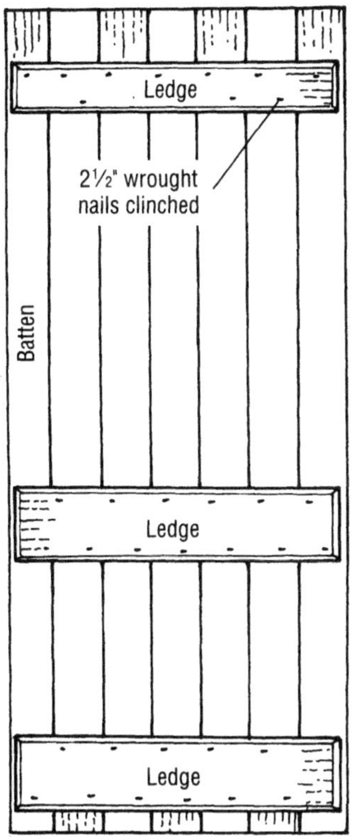

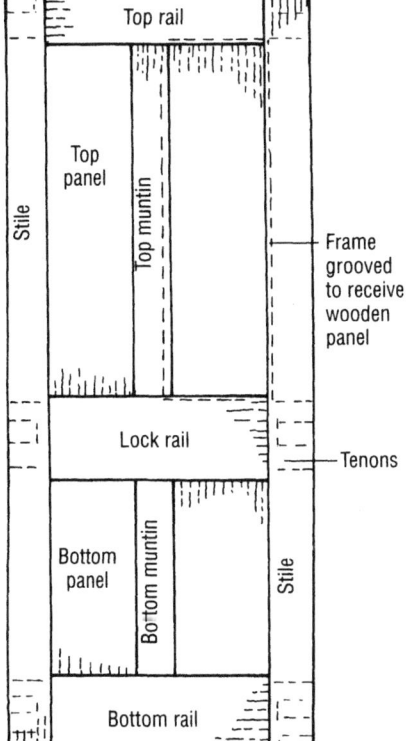

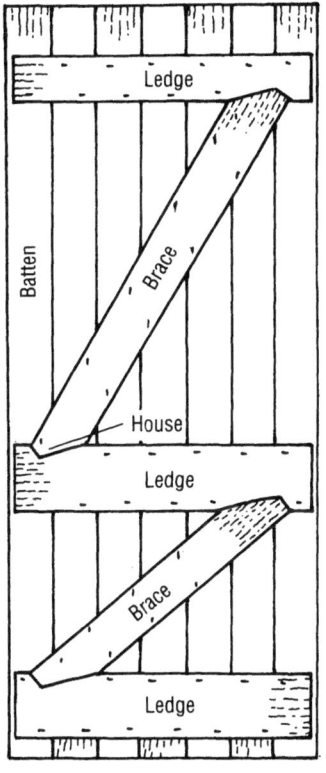

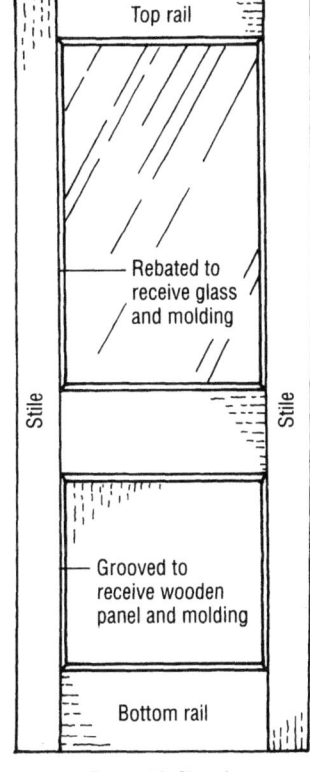

Four archetypal doors

Top left: *The traditional batten and ledged door consists of a number of vertical members termed* batten*s, or boards, that are fixed to a number of horizontal members termed* ledges*. The battens are secured by wrought iron nails that are driven through from the face of the battens—through the battens and then through the ledges—where they are then clinched.*

Top right: *The battened ledged and braced door—sometimes referred to simply as the ledged and braced door—is similar to the battened and ledged door, except that there are diagonal boards set between the ledges. The function of the brace boards is to prevent the door from drooping at the nose. For full effect, the brace boards should always spring up from the hinge batten.*

Bottom left: *Framed and paneled doors are generally described by the number of panels they contain. The two main vertical members are termed* stiles*, the horizontal members are* rails*, and the board set within the frame are* panels*. Although the panels are contained by the frame, they are free to move.*

Bottom right: *Sash doors are paneled doors that have one or more glazed panels set above the lock rail.*

wind, rain, and sun? If so, does it need to be fitted with weather bars and rain drips?

• How do you want the door to look—does it have to be prestigious, or do you want to copy an existing period or style?

• Does the door need to be fitted with a knocker and/or a number?

• Does it need to be glazed to allow extra light into a dark area, such as a hallway?

Once you have answered these questions, you will have a much clearer notion of your requirements, and you will be able to make an informed choice.

Anatomy of the Door and Door Joints

Being mindful that there are hundreds of years of door-making tradition behind us—with all manner of woodworking tools, techniques, countries of origin, designs, and procedures—the best way to start is to familiarize yourself with the parts of the door and the various door-making joints.

Fittings and fixtures are all the functional and decorative items that are fixed to the finished door: doorplate, hinges, knocker, latch, and so on.

Batten door

A common name for braced and ledged and for ledged, framed, and braced doors, meaning just about any simple door in which one side is made from battens. Wrought iron nails are driven in from the face side of the door and clinched over on the battens and braces. The pattern of nails indicates the underlying ledge and brace structure.

Batten

Any converted timber that is between 1½ and 2 inches thick and less than 9 inches wide. Mostly the term refers to the vertical timbers on the outside of ledged doors. However, a batten has come to mean any small, thin board about 1 inch thick.

Match boards

Battens and boards worked with tongues and grooves on the edges, so that they can be slotted together.

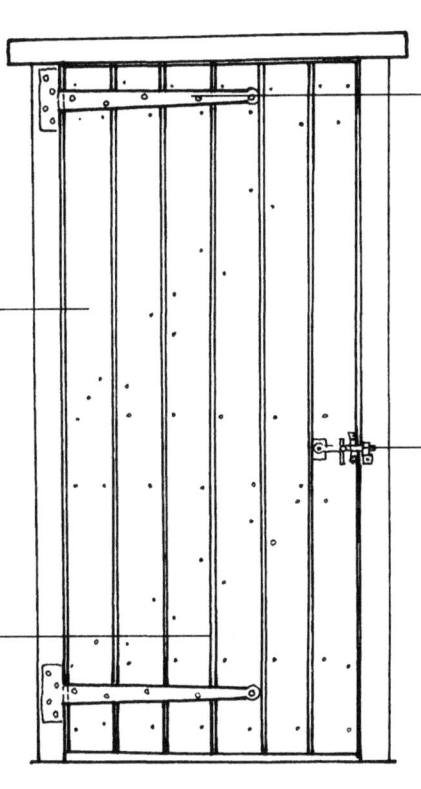

Strap hinge

A type of hinge with a long horizontal part that is screwed to the face of the ledge, usually fitted to heavy ledged and braced doors. Hinges of more or less the same character but designed to be fitted to lighter doors are known as cross-garnet and tee hinges.

Norfolk latch

A traditional metal handle and latch used on ledged and braced doors. The handle is gripped and the latch is pressed down with the thumb. In England, various counties had their own types of latches—an Essex latch, a Suffolk latch, and so on.

Framed and Paneled Door

Stiles, rails, and panels make up the *frame* of a framed and paneled door. *Stiles* are the outer vertical components. *Rails* run horizontally from stile to stile. *Panels* are the square or rectangular components contained by the frame.

Pediment

A pediment is a decorative crown above a door. All the features that make up the crown are inspired by classical architectural porticoes and gables. Variously, a pediment might be broken, open, segmented, or scrolled.

A broken pediment with an ornamental finial at the center.

An opened segmented pediment with carved swags. The term *open segmented* refers to the break in the arch; *swags* relate to the garlands of flowers.

Shutting stile

The stile on which the handle and lock are fitted—the opposite side of the hanging stile or lock stile.

Fingerplate

An ornamental plate fixed on or around the door handle or lock that saves the surface of the door from wear and dirt. These can be made from pressed steel, enameled copper, brass, glass, ceramic, and so on. Prestigious doors tend to have very decorative fingerplates.

Scutcheon or escutcheon

A little plate fitted over the keyhole.

Frieze

A decorative embellishment over the doorway—above the architrave but below the cornice. It might be carved from wood or stone or be made of plaster.

Top rail

The topmost horizontal rail of a framed and paneled door.

Architrave

See page 13.

Lock rail

The horizontal rail of a framed and paneled door on which the lock is fitted.

Flat panel

A panel without molding or a raised field.

Skirting

See page 14.

6 • *Doors*

Portico

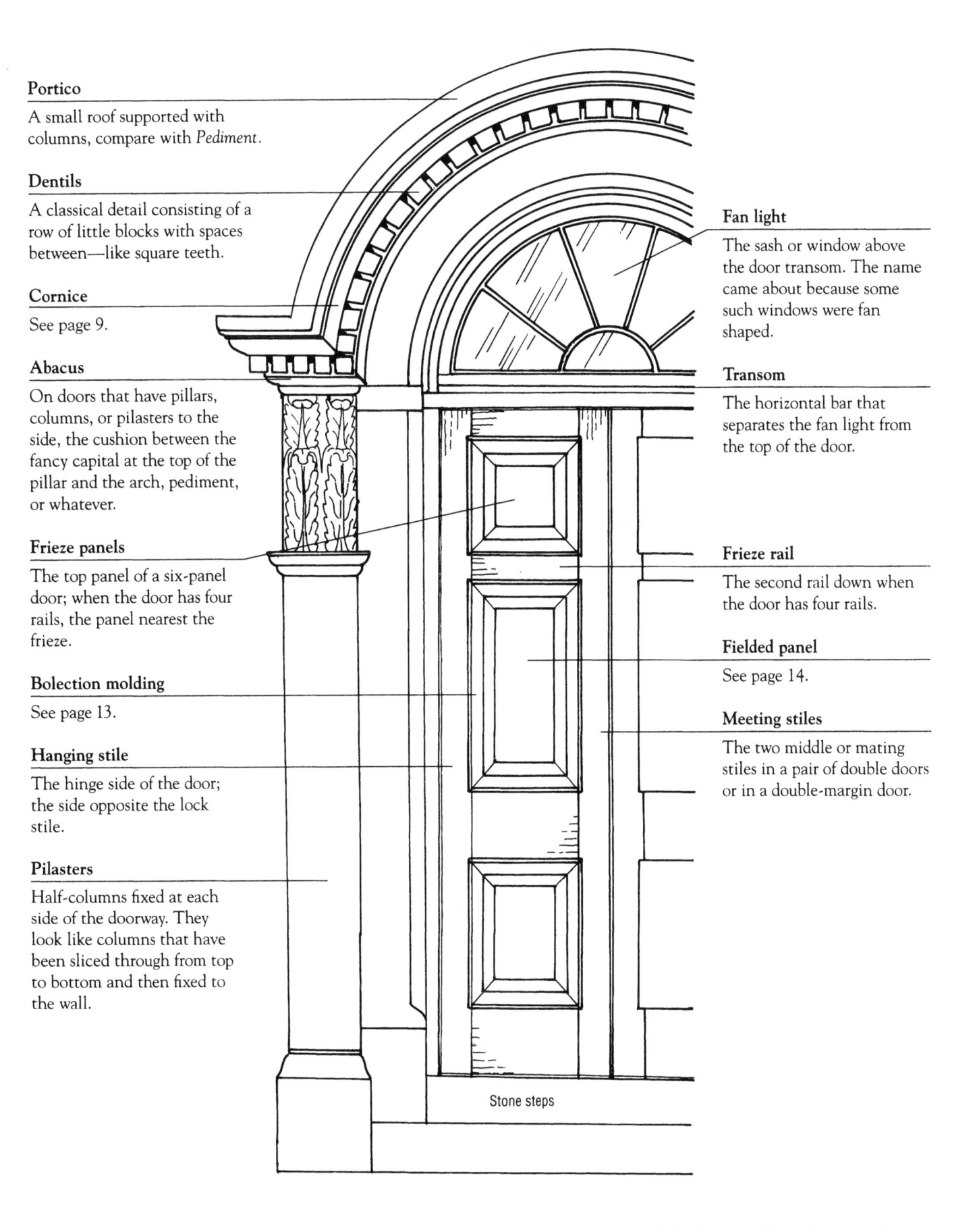

A small roof supported with columns, compare with *Pediment*.

Dentils

A classical detail consisting of a row of little blocks with spaces between—like square teeth.

Cornice

See page 9.

Abacus

On doors that have pillars, columns, or pilasters to the side, the cushion between the fancy capital at the top of the pillar and the arch, pediment, or whatever.

Frieze panels

The top panel of a six-panel door; when the door has four rails, the panel nearest the frieze.

Bolection molding

See page 13.

Hanging stile

The hinge side of the door; the side opposite the lock stile.

Pilasters

Half-columns fixed at each side of the doorway. They look like columns that have been sliced through from top to bottom and then fixed to the wall.

Fan light

The sash or window above the door transom. The name came about because some such windows were fan shaped.

Transom

The horizontal bar that separates the fan light from the top of the door.

Frieze rail

The second rail down when the door has four rails.

Fielded panel

See page 14.

Meeting stiles

The two middle or mating stiles in a pair of double doors or in a double-margin door.

Stone steps

Anatomy of the Door and Door Joints • 7

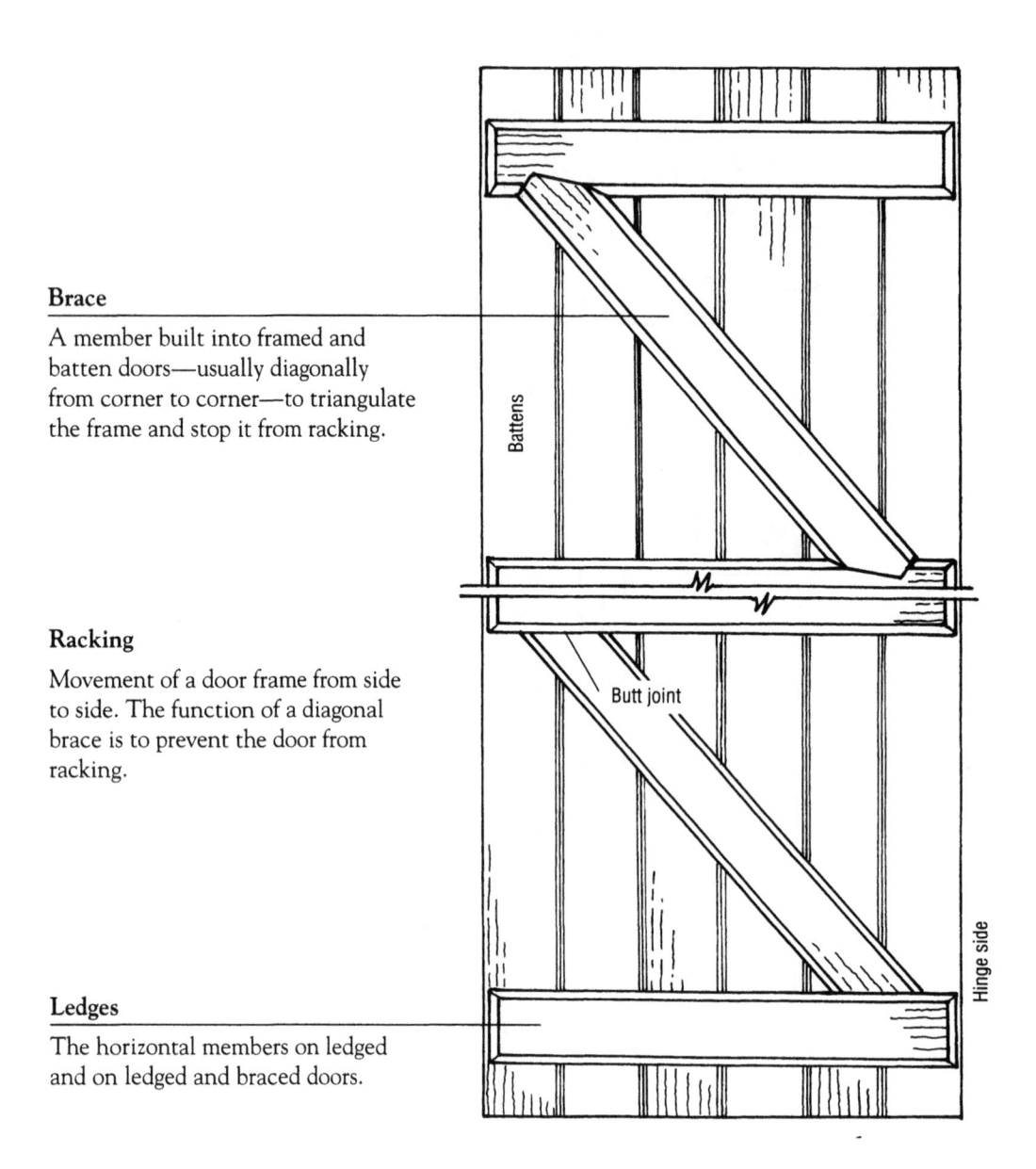

Brace

A member built into framed and batten doors—usually diagonally from corner to corner—to triangulate the frame and stop it from racking.

Racking

Movement of a door frame from side to side. The function of a diagonal brace is to prevent the door from racking.

Ledges

The horizontal members on ledged and on ledged and braced doors.

Battens

Butt joint

Hinge side

8 • *Doors*

Cornice

The projecting molding at the top of a frame or column that is built up from a number of sections and secondary moldings.

Bead

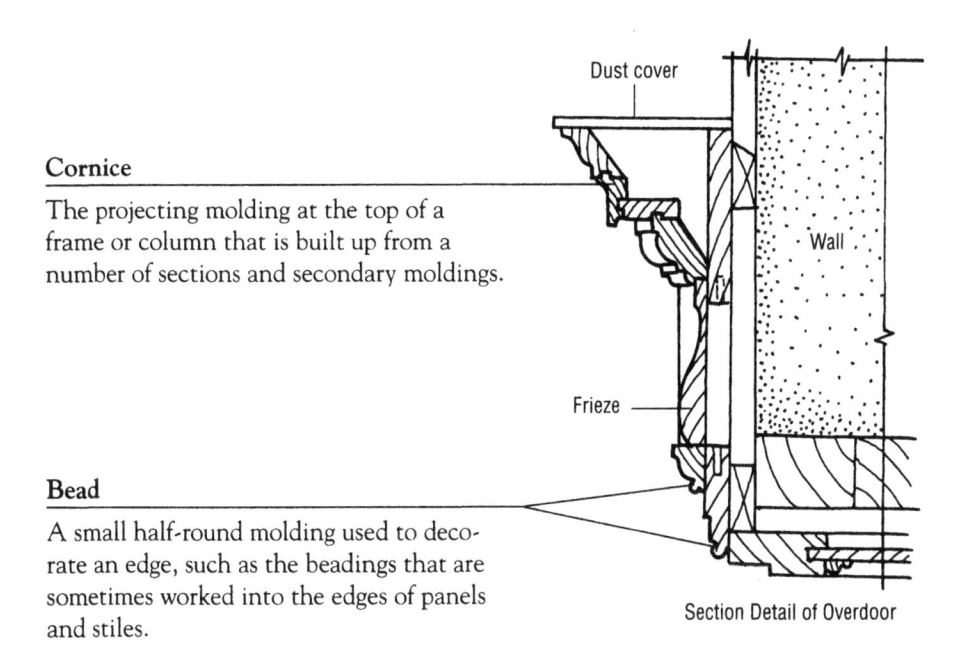

A small half-round molding used to decorate an edge, such as the beadings that are sometimes worked into the edges of panels and stiles.

Section Detail of Overdoor

Door frame

The wooden frame that carries the door and to which the door is hinged. External door frames are one piece, with the various steps, rabbets, grooves, and stops all being cut from a single large section; internal door frames are made from a number of small component parts that are variously nailed, jointed, and screwed to one another. The door frame of an exterior door is made of solid wood and fits into a recess formed in the wall. It consists of vertical members called *posts*, or *jambs*, that are tenoned into a horizontal member called the *head*. Unlike framed and paneled doors, in which the horns are cut off, the horns are left in place and built into the wall.

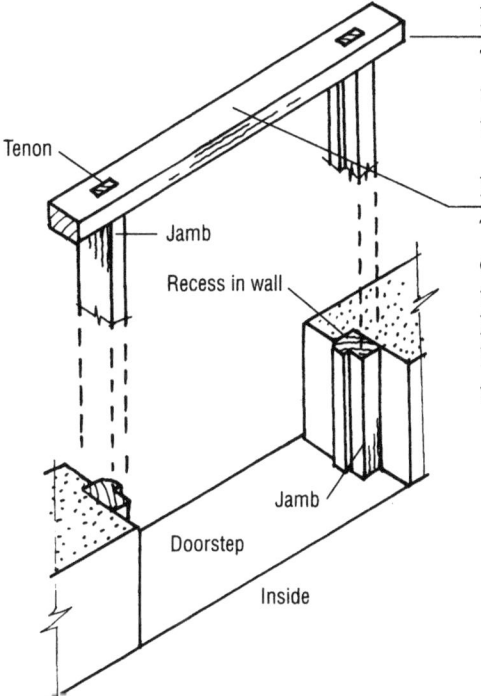

Horns

The ends of the stiles that need to be cut back; also, the ends of the head of a solid door frame.

Head

The top horizontal member of an exterior door frame. The posts of the frame are tenoned into the head in such a way that the horns left on the head are built into the wall.

Anatomy of the Door and Door Joints • 9

Jamb

The opening in which the door is fitted—specifically, the side of the opening.

Jamb lining

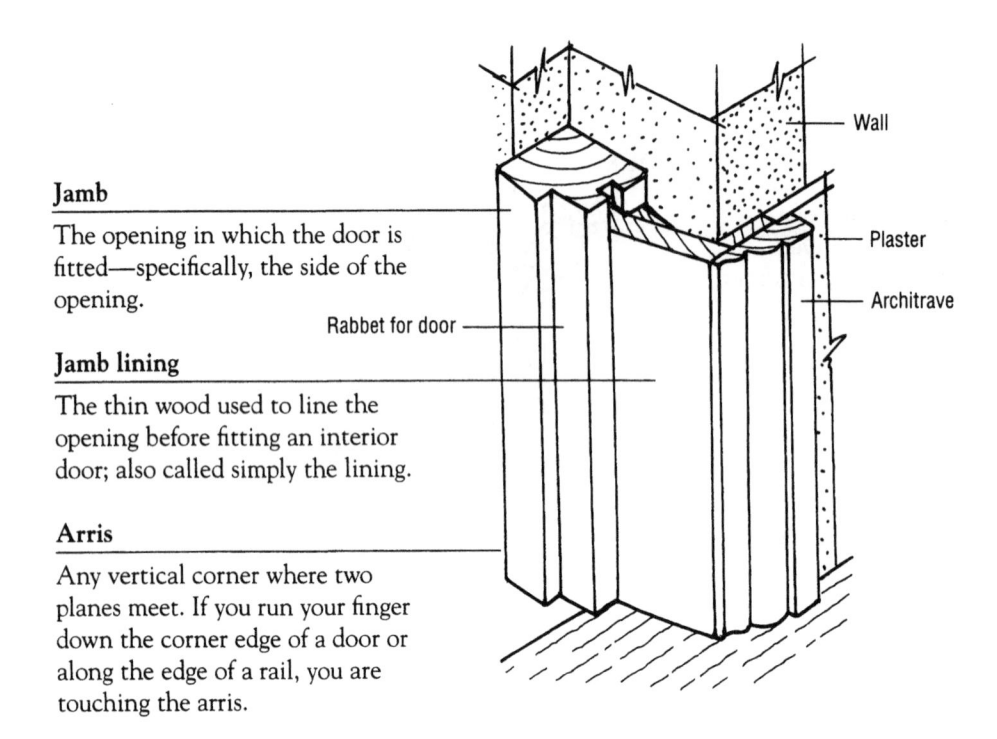

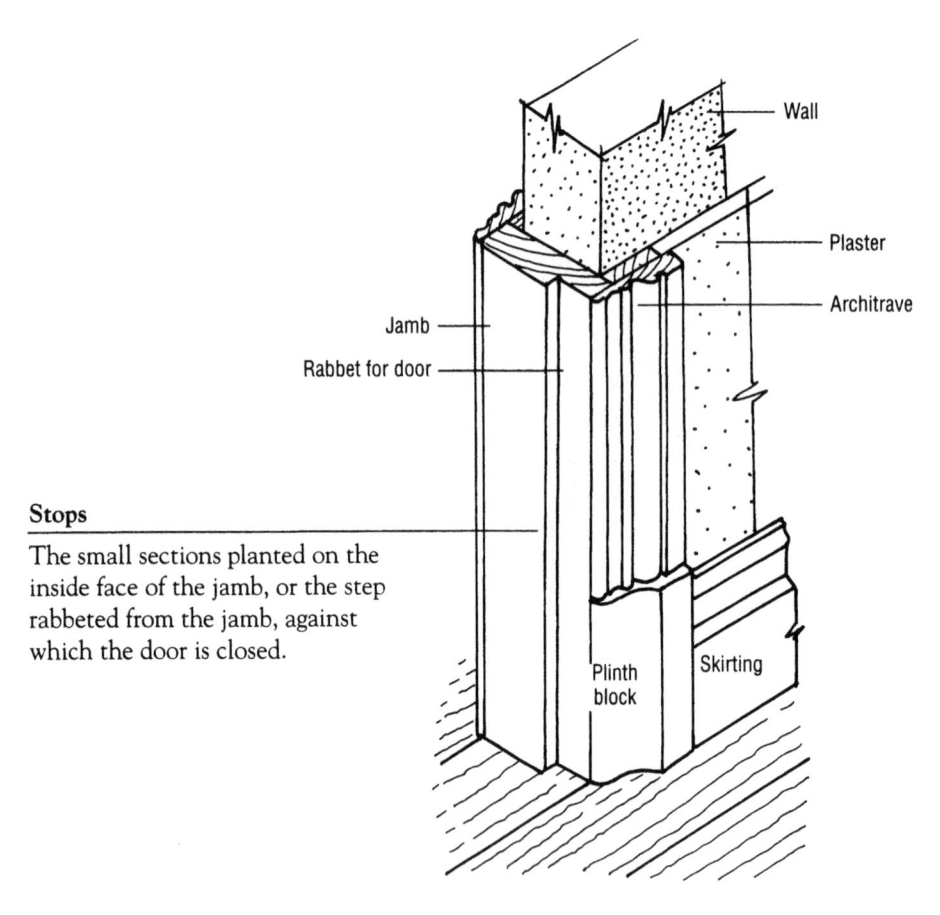

The thin wood used to line the opening before fitting an interior door; also called simply the lining.

Arris

Any vertical corner where two planes meet. If you run your finger down the corner edge of a door or along the edge of a rail, you are touching the arris.

Stops

The small sections planted on the inside face of the jamb, or the step rabbeted from the jamb, against which the door is closed.

Door linings

The woodwork that makes up an interior door frame. Sometimes the linings are termed *casings*. The complexity of linings or casings has to do with the thickness of the wall and the quality of the work. For example, a thin stud wall might have plain linings—no more than a batten with a nailed strip to act as a stop—whereas a thick wall might have double-rabbeted linings made from a frame with stiles, rails, and panels. With double-rabbeted linings, the stiles are rabbeted to become the stop or jamb.

Top: An exterior doorway viewed from inside the building, showing a solid frame fitted into a recess, with the lining components filling the gap between the frame and the inside face of the wall. Note the beautiful way the doorjamb frame is grooved to take the jamb lining.

Bottom: An interior door frame viewed from the door opening side, showing the way the architrave molding, the skirting, and the plinth block all come together. In this instance, the architrave would be fitted to the plinth block by means of a stopped housing and secret screws. Note that with high-quality work, the doorstop is rabbeted from the solid, rather than formed with a planted batten.

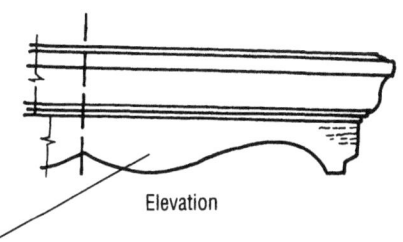

Cyma curve

A molding or profile that is S-shaped in cross section or form.

Apron rail

The middle rail of a door, when there is an additional molding on the upper edge. The function of the apron is to give design emphasis and to divert water from the bottom panel. Apron rail or molding and ornamental detail fitted on the top edge of the locking rail offer protection to the rail and panel. Note the way the apron is rabbeted into the rail.

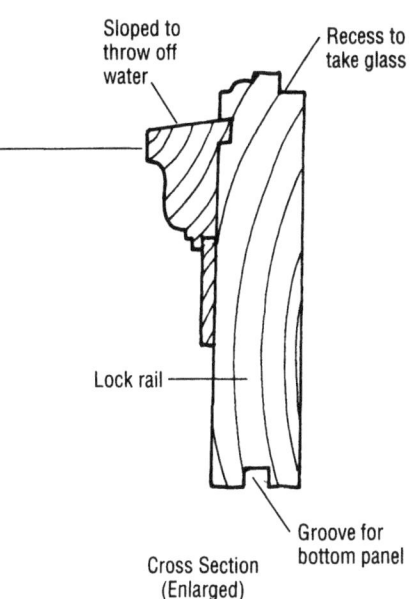

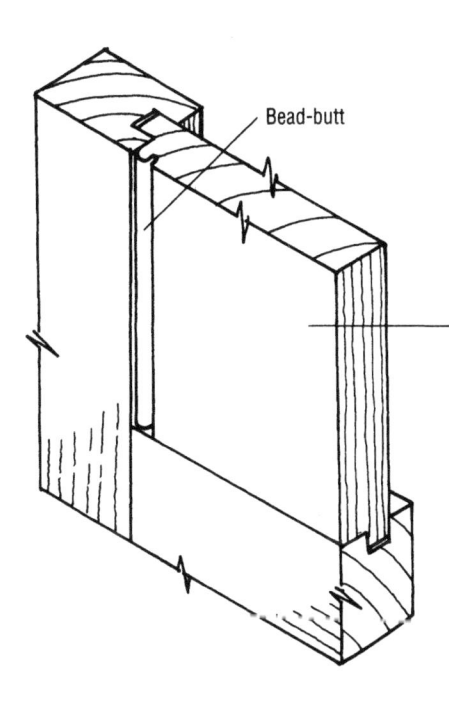

Flush panel

A panel that is level with the surrounding frame. This flush bead-butt panel shows how the bead is worked on the vertical sides of the panel. With the face of the panel finishing flush with the frame, the top surface of the bead is set about 1/16 to 1/32 inch lower than the frame. The panel is about two-thirds the thickness of the frame.

Anatomy of the Door and Door Joints • 11

Chamfer

If you remove the arris—as on the corner of a frame—the resulting beveled surface is the chamfer. Here is a selection of traditional chamfer stops. The chamfer is usually worked at an angle of 45 degrees to the face, where it travels along the length of the wood to stop a short distance from the end. The detail at the end of the chamfer is termed the *stop*. Fancy stops—such as the broached and molded stops—are carved with a chisel and gouge, whereas the plain stop is worked with a chisel and a template.

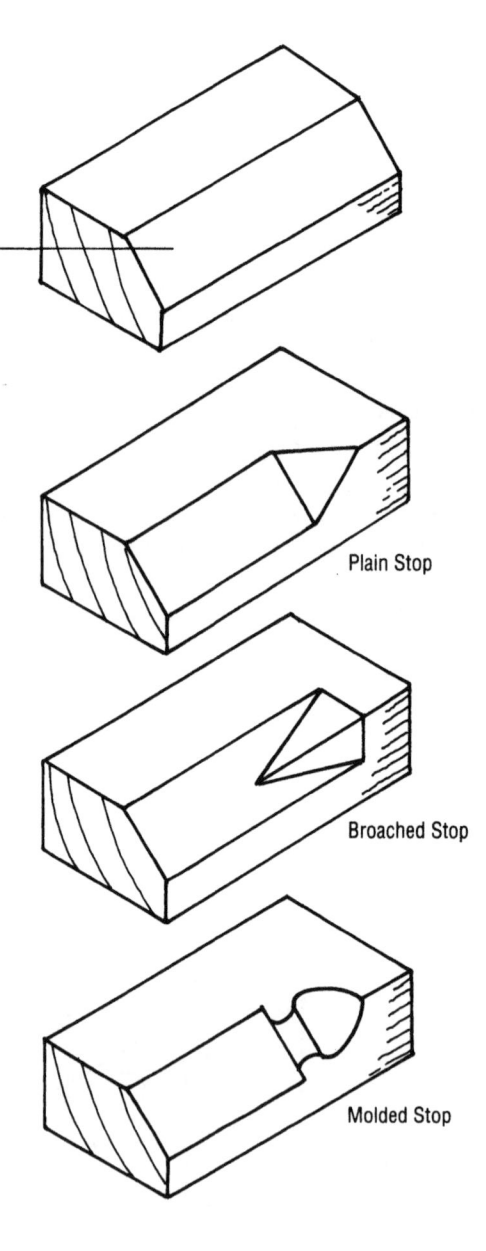

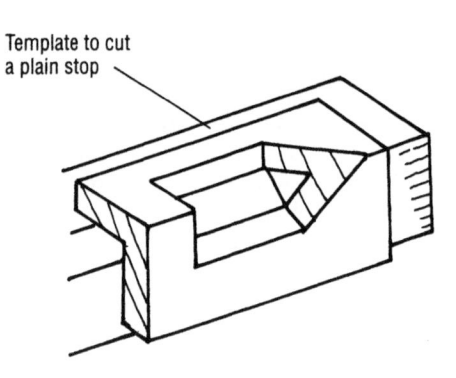

Plain Stop

Broached Stop

Molded Stop

Template to cut a plain stop

12 • *Doors*

Bolection molding

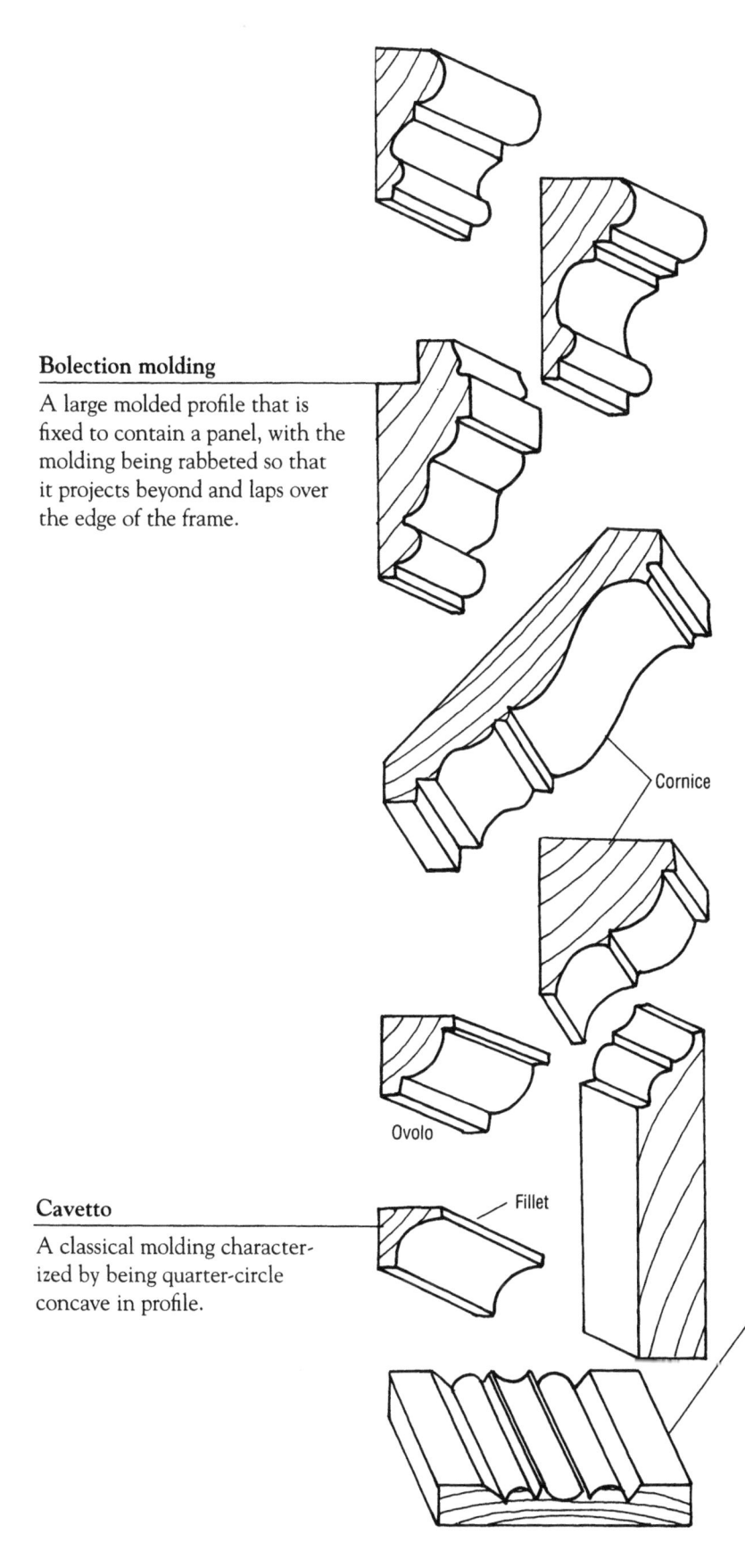

A large molded profile that is fixed to contain a panel, with the molding being rabbeted so that it projects beyond and laps over the edge of the frame.

Cavetto

A classical molding characterized by being quarter-circle concave in profile.

Molding

Lengths of wood that are shaped in section so that the shape runs along the length. Moldings are used both functionally and decoratively. For example, a molding might be used to hold a panel in its frame, to throw off water, or as a decorative light-catching feature. Moldings can be made with wooden molding planes, metal combination planes, a router, or a combination of sawing and carving techniques. Most of the traditional wooden moldings used on doors take their names from classical Greek and Roman originals, which were forms found in stone. Small moldings are best mitered at the angles and planted on the frame. Larger moldings over 1 inch wide need to be mitered, tongued, and screwed at the angles. Bolection moldings are secured either by running nails through the molding into the frame or by driving screws through the panel and then the molding. The molding needs to be fixed to either the frame or the panel—never to both. The moldings that make the cornice are generally secured by being tongued and grooved.

Architrave

The molding that runs around the door frame—on the inside of exterior door frames and on both sides of interior frames—and bridges the joint between the doorjamb lining and the plaster. As well as being decorative, it protects the edge of the plaster and conceals any movement of the jamb lining.

Anatomy of the Door and Door Joints • 13

Planted molding

A molding that has been nailed or glued to the surface rather than one that is an integral part of the surrounding woodwork.

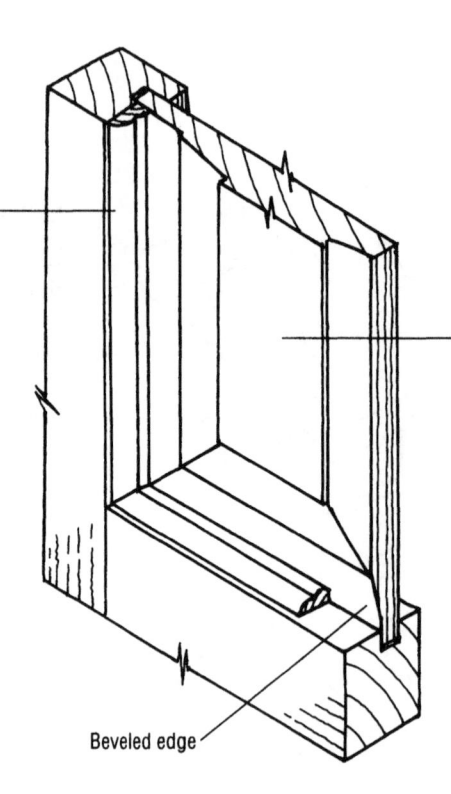

Fielded panel

A panel that has the central area stepped up from the sides so that the center is at a higher level than the sides; also known as a raised panel. A classical fielded panel is best defined as a panel that is thicker in the middle than at the edges, so that the middle area is like a plateau that sits above the general surface. The edges of the panel are reduced in thickness so that they fit into a circumscribing groove.

Beveled edge

Skirting

The board that runs around the bottom of the wall to meet the door plinth. The top of the board is usually molded. The skirting is generally nailed to the ground or grounds—that is, battens plugged to the wall—with the nails being punched below the surface and the whole works being filled and painted. Note that with this arrangement there is a useful space between the back of the skirting and the area of unplastered wall—good for hiding electric cables and pipes. In top-quality work, the skirting is tongued into the floor.

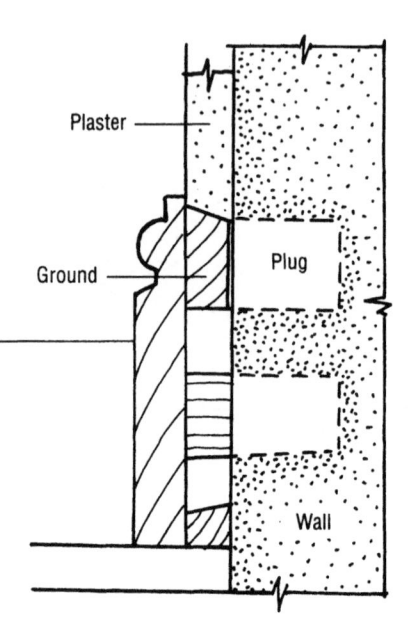

Section Through Skirting and Wall

14 • *Doors*

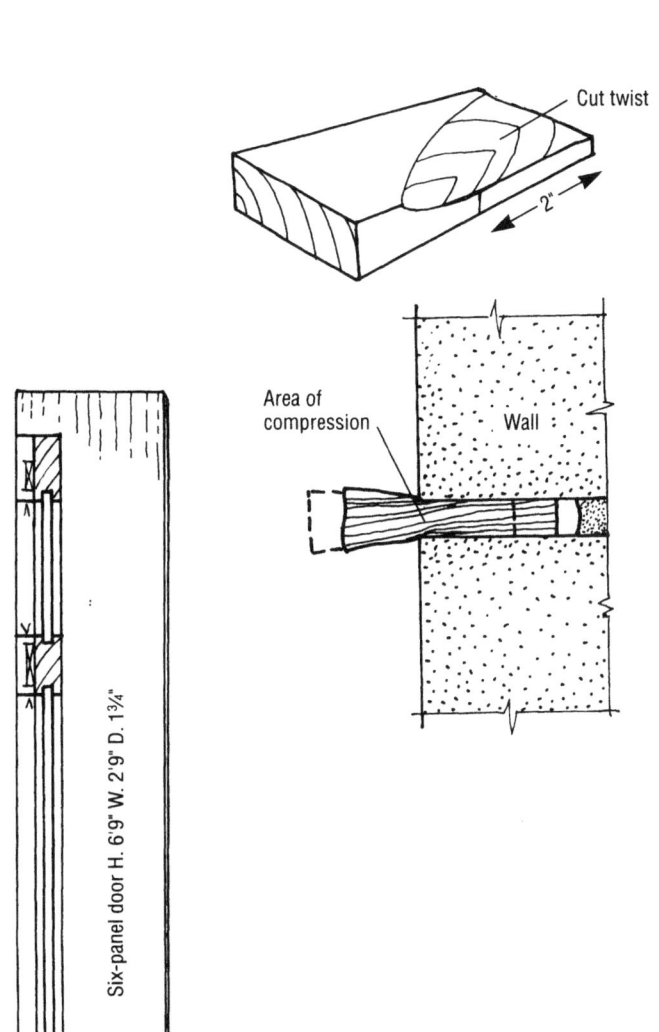

Wooden plugs

Small wedges of wood driven into the joints of a brick or stone wall on either side of the door opening to provide attachment points for the door frame and linings.

Top: The wedge is cut from dry, straight-grained wood with an ax or heavy knife, so that there is a little twist cut on each surface. The idea is that as the wedge is banged home, it slowly twists and screws into place.

Bottom: Cross section through the wall and the plug. At least 2 inches of plug must be in the wall.

Rod

A board on which the various sections are drawn out full size. A good door rod can be used to transfer the dimensions and forms directly to the wood being used. This rod shows all the details and measurements for a six-panel door drawn full size. The idea is that you can read off and transfer the measurements and sections directly to the wood and thus make swift and multiple components. Traditionally, door makers built up whole collections of such rods, each carefully labeled and identified.

Anatomy of the Door and Door Joints • 15

DOOR-MAKING JOINTS AND JOINT COMPONENTS

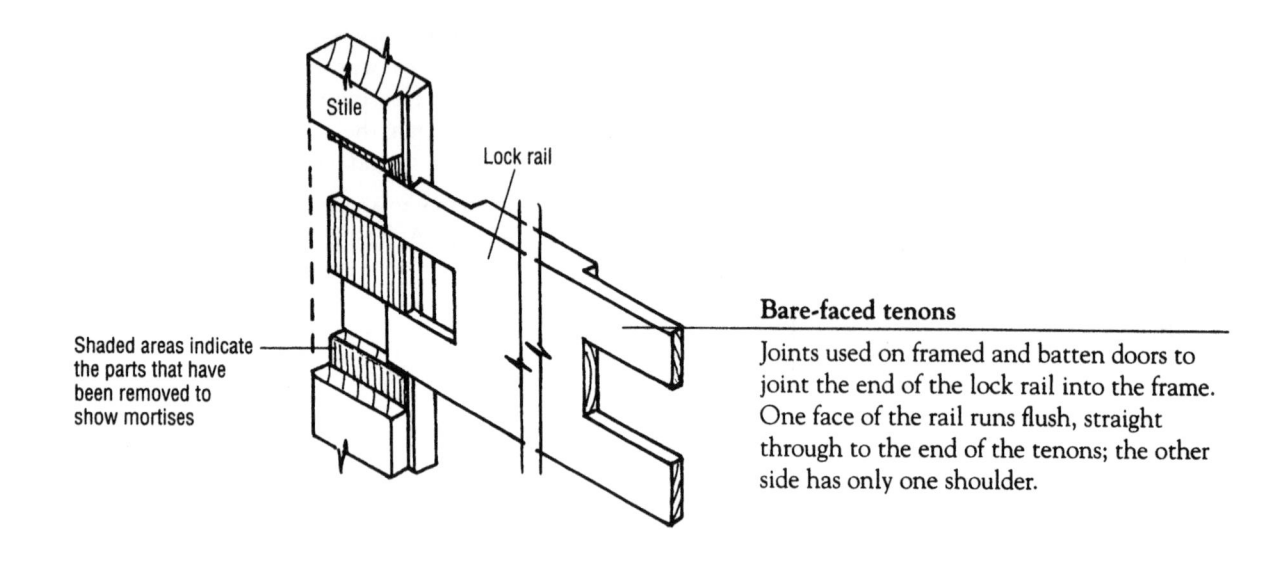

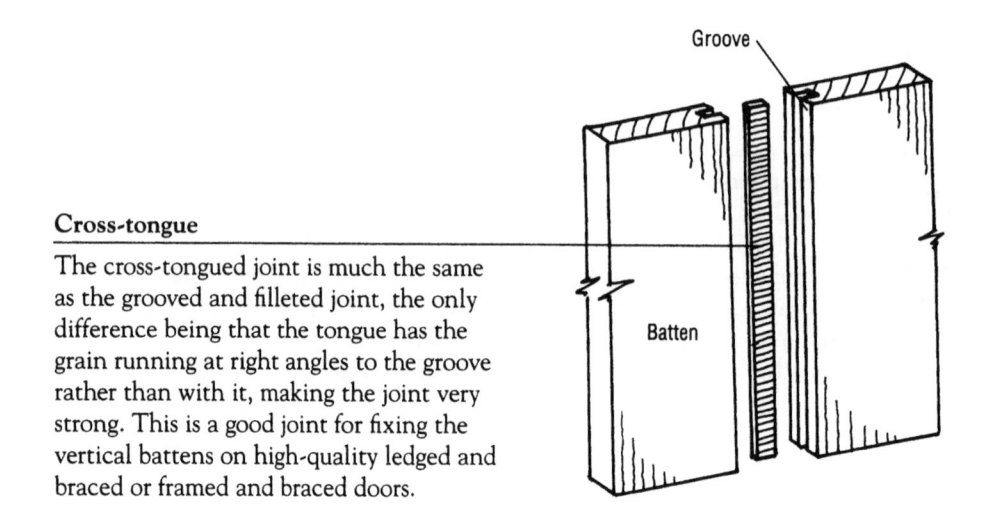

Bare-faced tenons

Joints used on framed and batten doors to joint the end of the lock rail into the frame. One face of the rail runs flush, straight through to the end of the tenons; the other side has only one shoulder.

Cross-tongue

The cross-tongued joint is much the same as the grooved and filleted joint, the only difference being that the tongue has the grain running at right angles to the groove rather than with it, making the joint very strong. This is a good joint for fixing the vertical battens on high-quality ledged and braced or framed and braced doors.

Dado

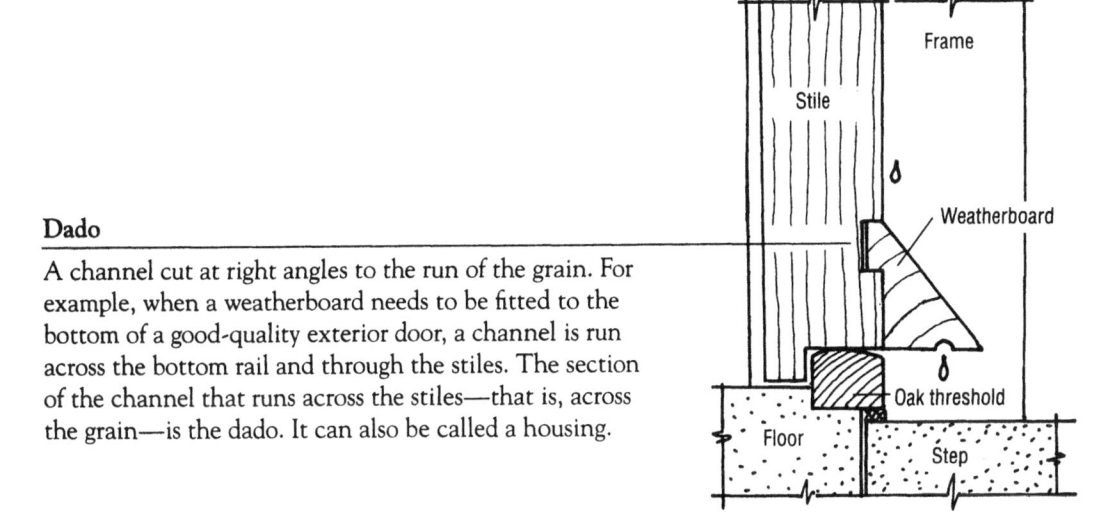

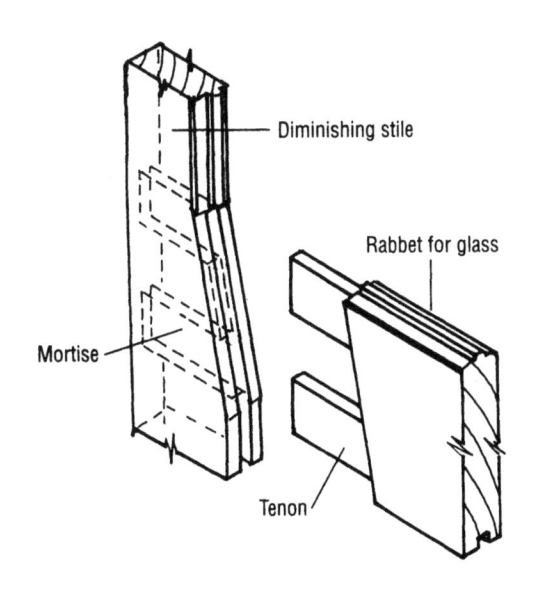

A channel cut at right angles to the run of the grain. For example, when a weatherboard needs to be fitted to the bottom of a good-quality exterior door, a channel is run across the bottom rail and through the stiles. The section of the channel that runs across the stiles—that is, across the grain—is the dado. It can also be called a housing.

Diminished stile mortise and tenon

Stiles that are reduced in width above the level of the lock rail (in doors that are glazed above the rail) are termed *diminished stiles* or *gunstock stiles*. The difficulty with diminished stiles has to do with cutting the tenons and the shoulders on the lock rail.

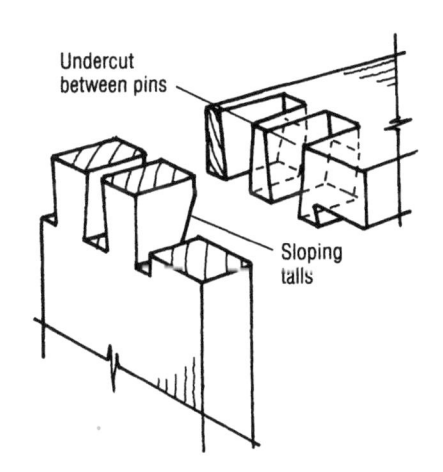

Door-frame dovetail

This joint is used on quality door jambs at the top corners. Not only is this joint stronger than the usual wedge and nail method and the rabbet method, but its design allows the frame to move without the joint appearing to open.

Double-paired tenons

A joint used on the end of the lock rail—on the lock side only—for extra-thick framed and paneled doors. There is some confusion as to the name; they've been variously described as paired double tenons, a double pair of single tenons, a pair of double tenons, and so on. Basically, there are four tenons with a shoulder running to all sides. The mortise is cut in the same way as for all door tenons—like a dovetail key, so that when the wedges are driven home from the outside of the frame, they convert the joint into a dovetail.

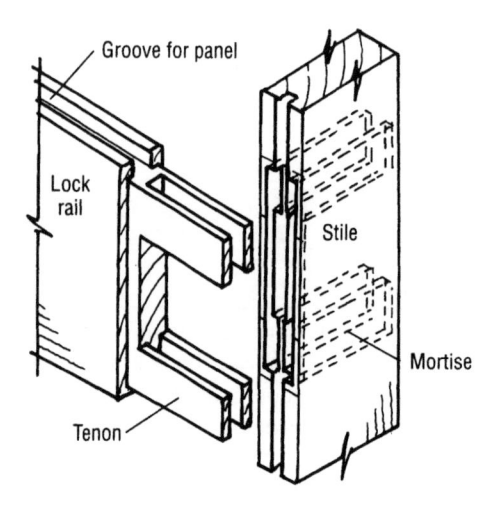

Folding wedges

When two narrow framed and paneled doors are put together to make double-margin doors, the two doors are fitted together with several pairs of folding wedges. The wedges are set in the mortises and banged together so that they slide past each other and tighten.

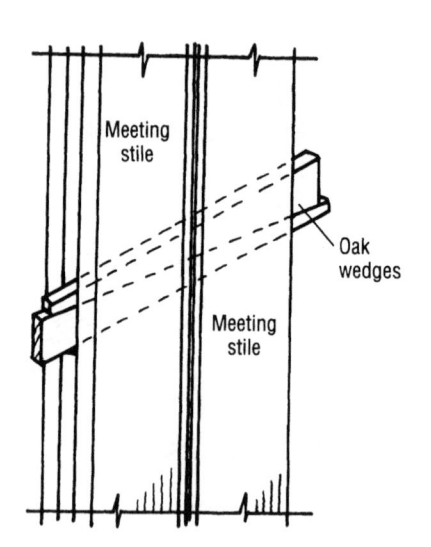

Foxtail wedging

The mortise is stopped about ½ inch short of the back edge of the stile and cut slightly dovetail in shape. The tenon has a saw cut ⅛ inch in from each side, into which a wedge is inserted. When the joint is put together and clamped, the wedges spread the tenon and lock it in the mortise. This joint is used on top-quality doors when you don't want to see the end-grain tenons penetrating the stiles.

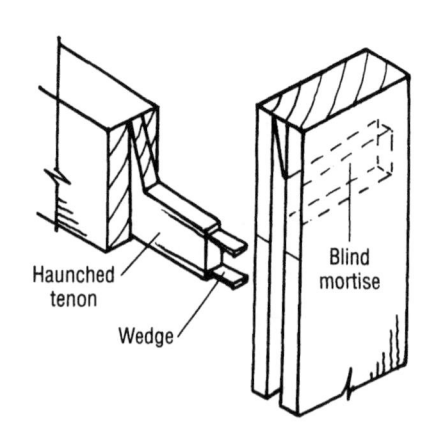

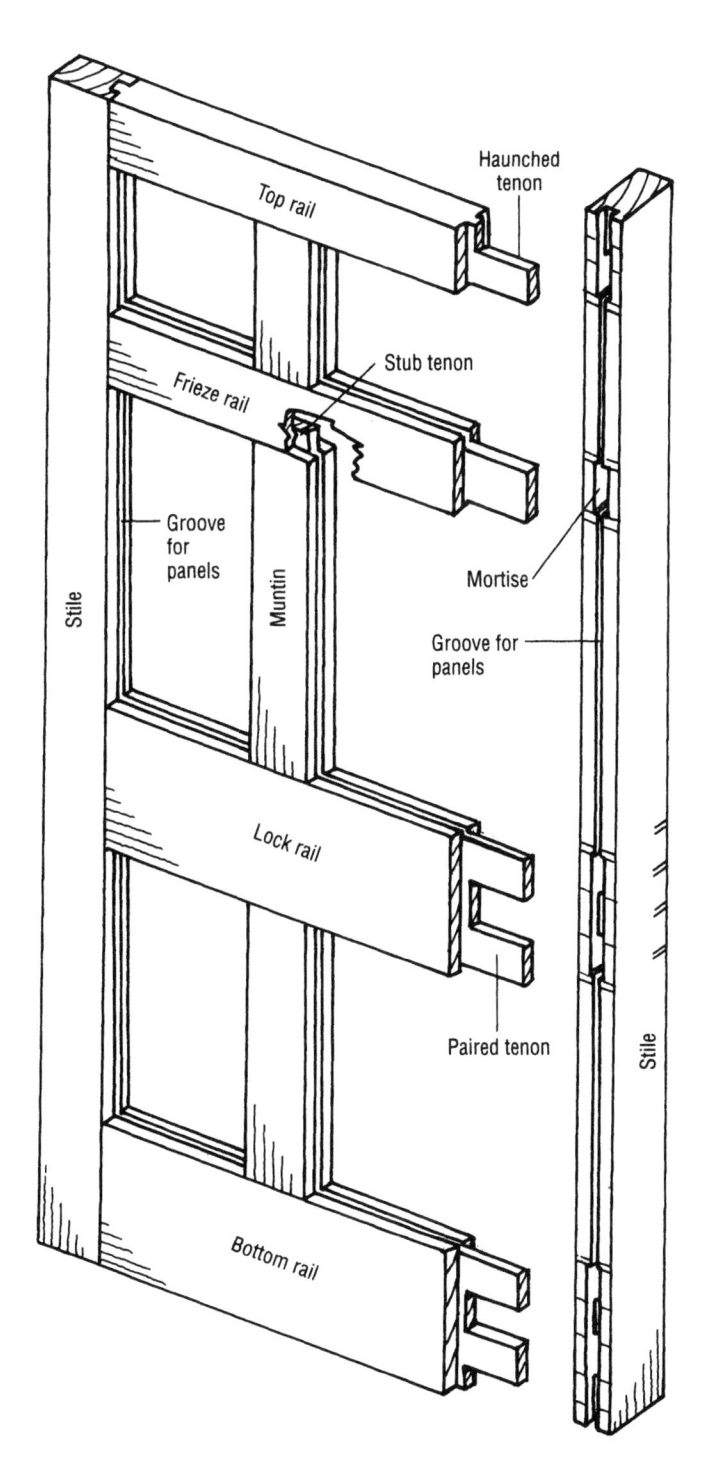

Framed-door joints

A frame is usually constructed with various mortise and tenon joints. The frame encloses a number of rectangular or other shaped pieces that are termed *panels*. The frame usually has grooves or rabbets on the inside edges, into which the panels are fitted. The frame completely encloses the panel to prevent it from twisting.

Groove

A narrow slot cut in the direction of the grain, designed to take a tongue, a feather strip, the edge of a panel, a rabbet, and so on.

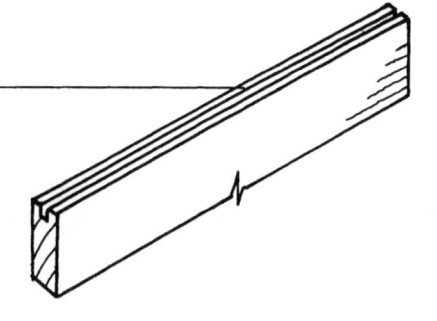

Haunch and haunching

The haunch is the short stump left when cutting the tenons and the haunching is the mating part of the mortise. The sum of the haunchings on the rails prevents the tenon from twisting and does away with the need for a break at the end of the stile. The short haunch adds great strength to the root of the tenon.

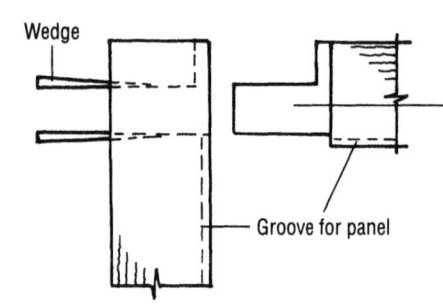

Haunched tenon

The joint used on the top and bottom door rails, when the edge of the rail needs to finish flush with the end of the stile. The haunch is used to prevent weakening of the end of the stile.

Miter

A joint in which two components are fitted at an angle other than a right angle. In the context of doors, the simple 45-degree miter—used for fitting panel moldings—is most common.

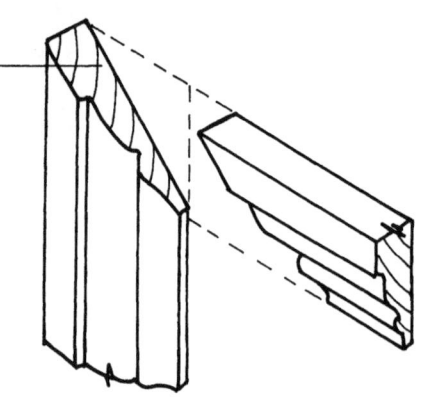

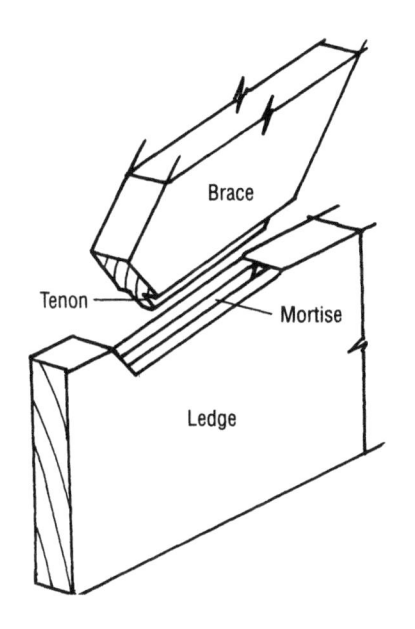

Oblique mortise and tenon

Much the same as a simple mortise and tenon joint, the only difference being that the tenon is more of a stub and the mortise is extended at an angle beyond the shoulder line, with the shoulder being cut at an angle. This joint is used on good-quality framed, ledged, and braced doors when the bottom end of the brace is jointed into the ledge.

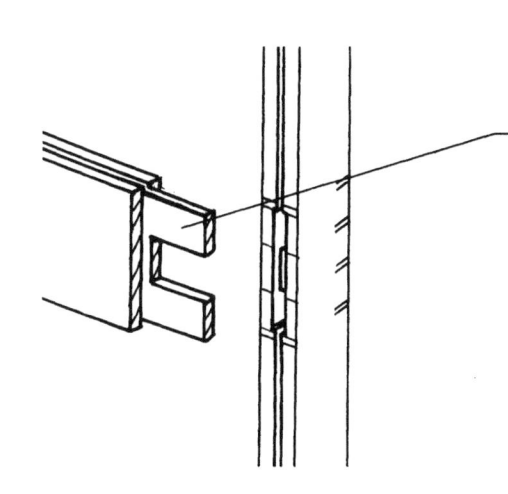

Paired tenons

The joint used on most panel doors for jointing the lock and the bottom rails to the frame; sometimes described as a pair of single tenons. The two tenons are usually haunched.

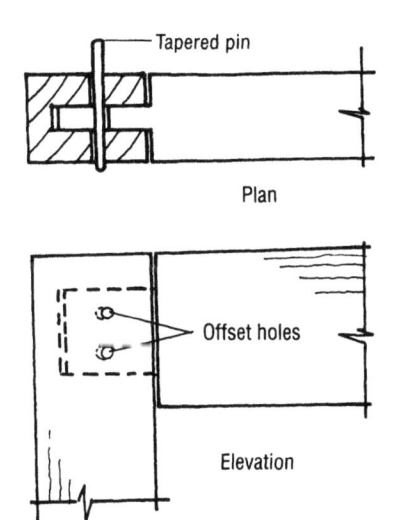

Pinned Tenon

A joint used on relatively lightweight framed doors, in which two dowels are run through the cheek—through the total joint. Although through dowels are both attractive and traditional, they sometimes fail when used on heavy doors.

Rabbet

A rectangular, sectioned step or recess in which the cut is made on the edge of the wood, such as the rabbet on the edge of a jamb into which the door closes, and the rabbet into which the boards and panels of some doors are fitted. The ability to cut rabbets is an important part of door making.

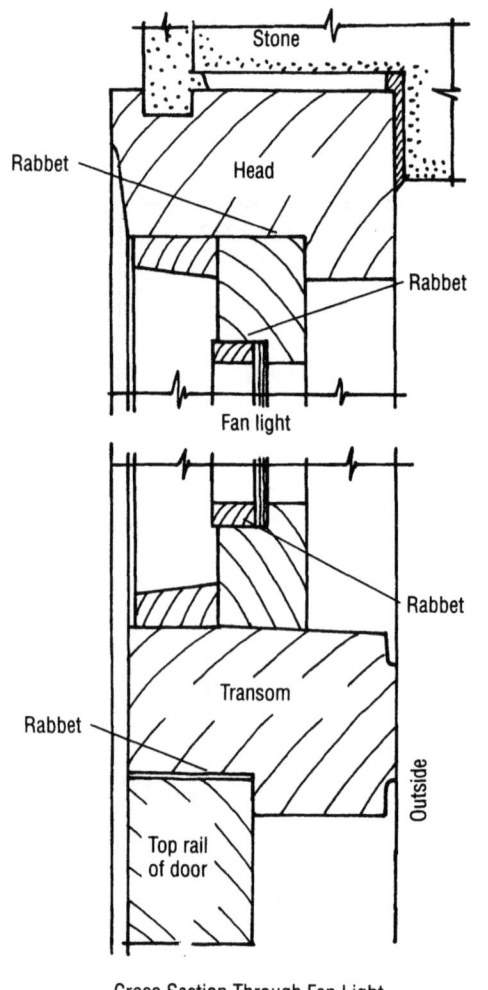

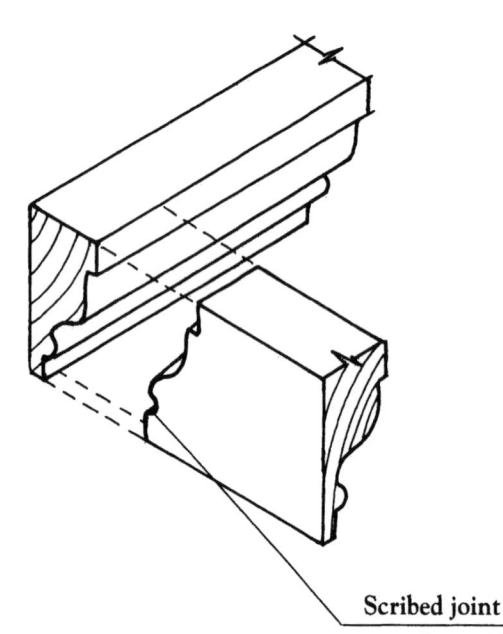

Cross Section Through Fan Light

Scribed joint

Joint formed by cutting the end of one molding to fit the profile of the other. When two lengths of the same molding meet at the corners—as when the architraves meet at the corners—the usual practice is to cut a mitered joint, but sometimes scribing is preferable.

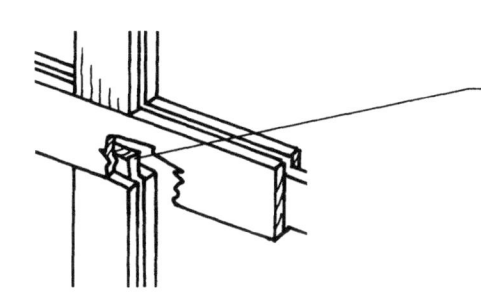

Stub tenon

A short tenon—usually cut on the ends of the muntin—that runs in a mortise cut in the rails.

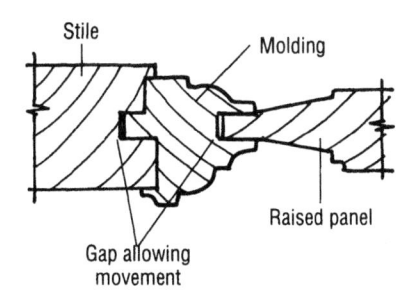

Tongued and grooved panel joint

A joint used to hold the panels in framed doors. The panel can be raised, so that the resultant tongued edge is let into a groove that runs on the inside of the frame, or it can be let into a grooved molding that is itself tongued into a groove that runs on the face of the frame. All these joints allow both the frame and the panel to move without cracking or opening.

Generic Door Types

LEDGED DOORS

Ledged doors are traditionally used for workshops, outbuildings, and lean-to sheds. They are the simplest form of doors. In essence, they consist of a series of vertical battens about ⅞ to 1 inch thick and 4 to 5 inches wide that are nailed to three horizontal boards or cross-ledges about 1½ inches thick and 6 to 9 inches wide. The battens may be square, chamfered, tongued and grooved, ploughed and tongued, or even cross-tongued. Although this type of door is an easy-to-make,

Batten and ledged door, showing a front or outside elevation, a section plan, and a side section.
Height: *6 feet 6 inches.*
Width: *2 feet 6 inches.*
Thickness of battens: *⅞ inch.*
Chamfered ledges: *5½ inches wide and 1 1/18 inch thick.*
This basic door is suitable for interior and exterior use in workshops and sheds and in period cottages. There are three ledges with tongued and grooved battens fixed with clinched nails. In this particular door, the strap hinges are fixed to the ledge face. The drip on the underside edge of the bottom rail suggests that the ledge side is designed to face the weather. Note also the three suggestions for edge jointing the battens.

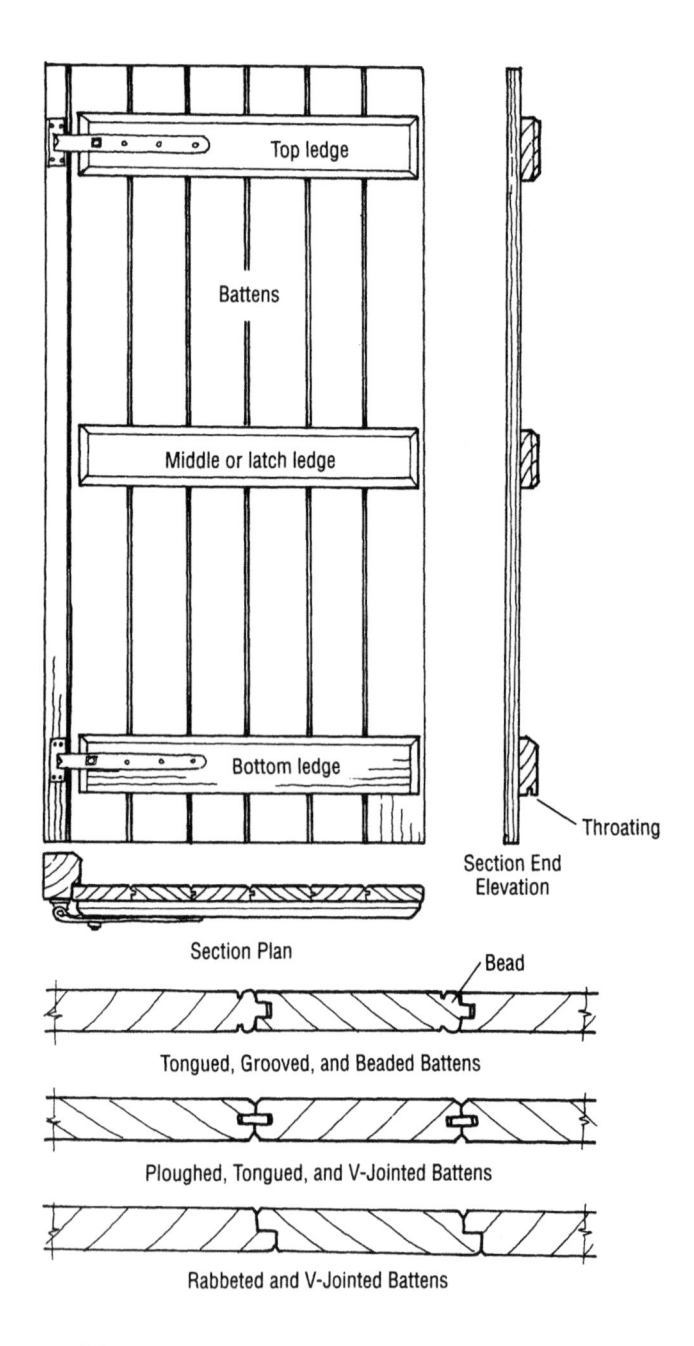

low-cost option, its main failing is that the weight of the door and the absence of supports tend to cause it to sag and warp. So this type of door may be perfect for the chicken shed or the outside toilet, but it wouldn't serve too well as the main door to a palatial mansion. However, some cottage doors constructed in this way—but built of 1½-inch-thick English oak, with only two vertical boards about 14 inches wide and the whole structure being held together with clinched forged nails rather than wire nails—are over 200 years old and still going strong.

LEDGED AND BRACED DOORS

The ledged and braced door—sometimes known as a Z-door—was almost certainly built in answer to the problem of sagging ledged doors. It is, in effect, a ledged door with the addition of diagonally set brace boards that slope upward from the hinged side of the door. The brace boards make for a much stronger door, in that they prevent the structure from dropping at the nose. There are usually two braces—one that runs from the bottom ledge through to the middle or latch ledge, and one that runs from the middle ledge through to the top ledge. Sometimes the braces are simply butted in place; with the more expensive option, the ends of the brace boards are let into the horizontal ledge with notches and/or oblique stub tenons. Doors of this type function perfectly well for workshops and stables and for cottage and porch doors where there is a need for country-type imagery. They are strong and they look good.

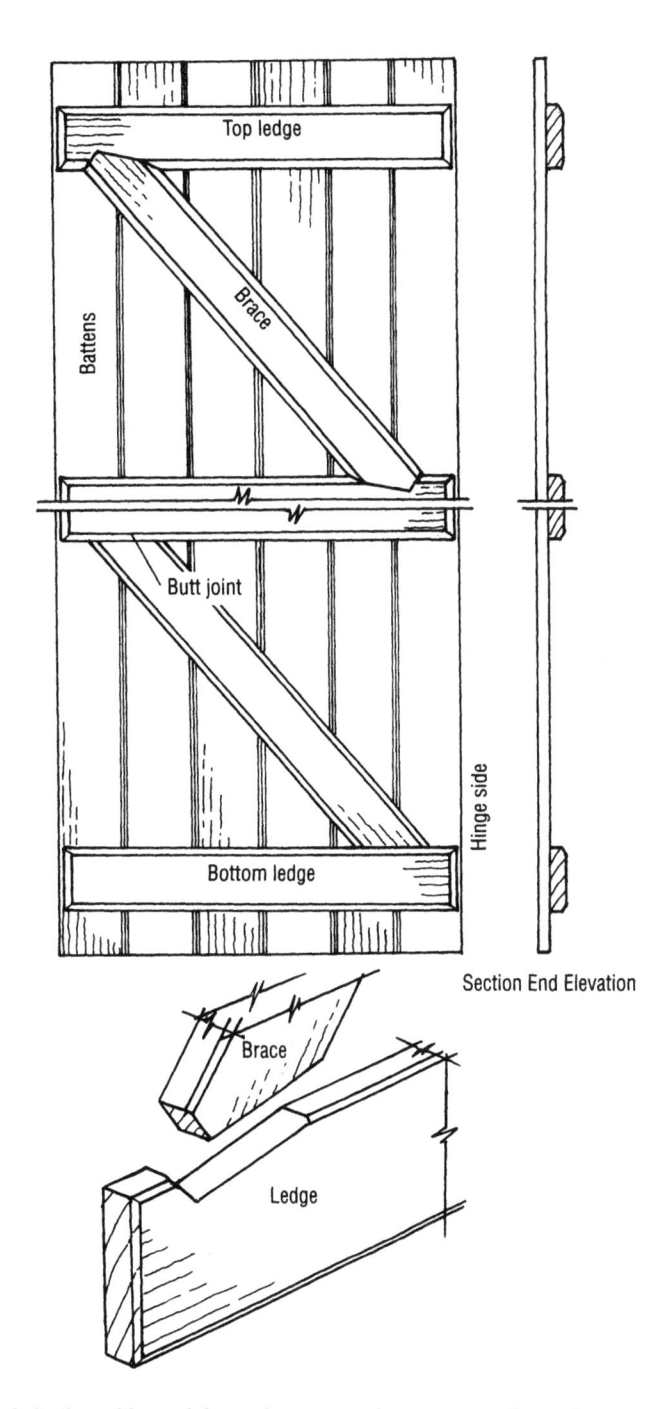

Batten, ledged, and braced door, showing a split two-option front elevation, a side section, and a brace-to-ledge detail.
Height: 6 feet 6 inches.
Width: 3 feet.
Thickness of battens: ⅞ inch.
Ledges and braces: 5½ inches wide and 1¼ inches thick.
Top: *The braces are butted against the ledges. Note how the ends of the ledges are set slightly back from the edges of the door, to simplify trimming the door back to width. See also how the braces spring up from the hinge side of the door.*
Bottom: *The braces are housed or notched in the ledges.*

Generic Door Types • 25

FRAMED AND LEDGED DOORS

The framed and ledged door is one step up from the simple ledged and braced door. The increased thickness of the stiles and ledges at 2 inches thick, with the ledges being tenoned into the stiles, makes it a strong, attractive door that is ideal for country cottage-type houses. In the example illustrated, the rails and battens are chamfered along their length, and the stiles are stop-chamfered to create a panel-like feature inside the door. With this type of door, the battens are either nailed directly over the whole outside of the frame or set within the frame. The battens were traditionally secured with wrought iron nails hammered in from the face and clinched at the back.

Framed and ledged door, showing an inside-house elevation, a section plan, a side section, and tenon joint details.
Height: *6 feet 4 inches.*
Width: *2 feet 4 inches.*
Thickness: *2 inches.*
This is a basic external door designed to be used where outside appearances are not important but there is a need for decoration on the inside face. Note how the top rail is haunch-tenoned into the stiles, with both the top stiles and the top rail being rabbeted to accept the battens. The thing that makes this door so special is the way that the ledges and the rails are chamfered—easy to do, and a very attractive feature. This type of door is best hung on good-sized cast iron butt hinges.

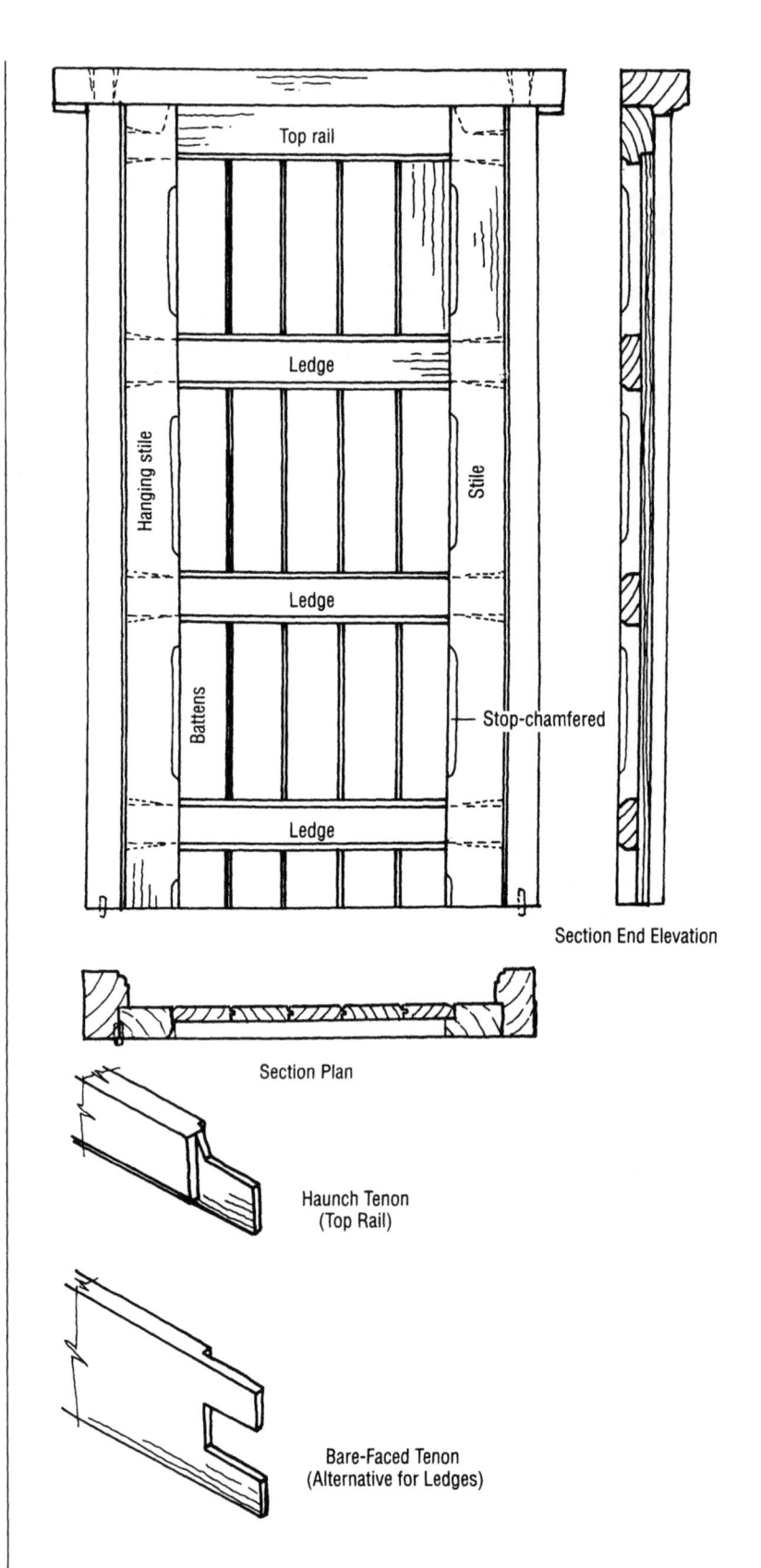

FRAMED, LEDGED, AND BRACED DOORS

Doors of this type are in essence ledged and braced doors with the addition of two stiles. The ends of the rails (ledges) are tenoned into the stiles, and the vertical battens are tongued into both the rails and the stiles. This has been described as a transitional door because it is one of the most complicated of the board and batten structures and one of the simplest of the framed and paneled structures. If you are looking for a structurally sound country door, this is the one to use. If you want something a bit fancier and lighter, the rails, braces, and stiles can all be stop-chamfered.

Framed, ledged, and braced door, showing an inside-house or back elevation and a side section.
Height: *6 feet 3 inches.*
Width: *2 feet 6 inches.*
Thickness: *2 inches.*
This is a first-choice door for workshops and outbuildings. It is strong and is good for both interior and exterior use. Note that there is a chamfer on all the edges of all the rails and braces and on the inside edges of the stiles. The braces are oblique stump-tenoned into the stiles and the ledges at the bottom ends but only notched into the ledges at the top ends. The reason is that if both ends of the brace were stump-tenoned, it would eventually push and open the joint between the rail and the stile. When putting doors of this character together, old-timers reckoned that for exterior doors (that is, doors that were going to get damp), it was better to forget the glue and to gloss paint all the mating faces directly before wedging and clamping up.

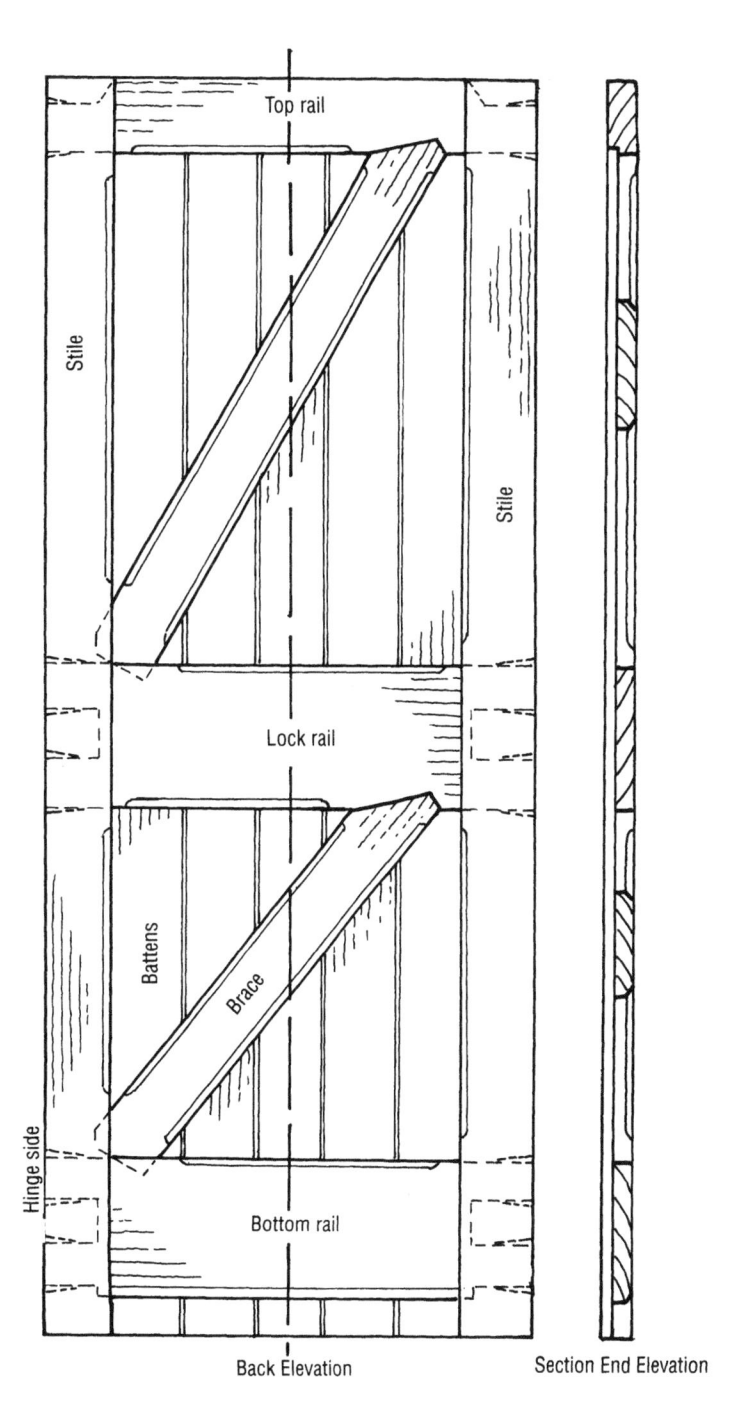

Back Elevation   Section End Elevation

The brace is stump-tenoned into the stile at the bottom end, and cut into the rail at the top end.

Generic Door Types • 27

Although there are regional variations, the top rail is generally about 5 inches wide, the lock rail and the bottom rail are about 9 inches wide, the stiles are about 4½ inches wide, and the braces are about 4 inches wide. Doors of this type are a good choice up to a width of about 33 inches; wider doors are often halved and hung as two narrow mirror-image doors.

Doors of the framed, ledged, and braced type have been used in all manner of situations—everything from workshops, outhouses, and porches to stables and Gothic churches. Although you might not have much use for a church-type door, it has always been fashionable in some country districts—especially in Britain and New

A pair of framed, ledged, and braced church doors, showing a front or outside elevation.
Height: *12 feet 4½ inches.*
Total width: *5 feet 6 inches.*
Thickness: *4 inches.*
If you are going to build a church or a chapel or you are simply looking for a pair of fantasy doors for your neo-Gothic house, these are the doors for you. They are framed with ledges and braces, with the face being battened and the whole works being held together with hand-forged hinges. The important thing here is that the hinges are structurally necessary. They are bolted through to the frame, ledges, and braces so that they help support and spread the total weight of the door. The wood should be oak—Gothic, church, and oak are the perfect combination. But be mindful that the tannin in oak eats iron, so have the door pegged up—no iron nails or screws.

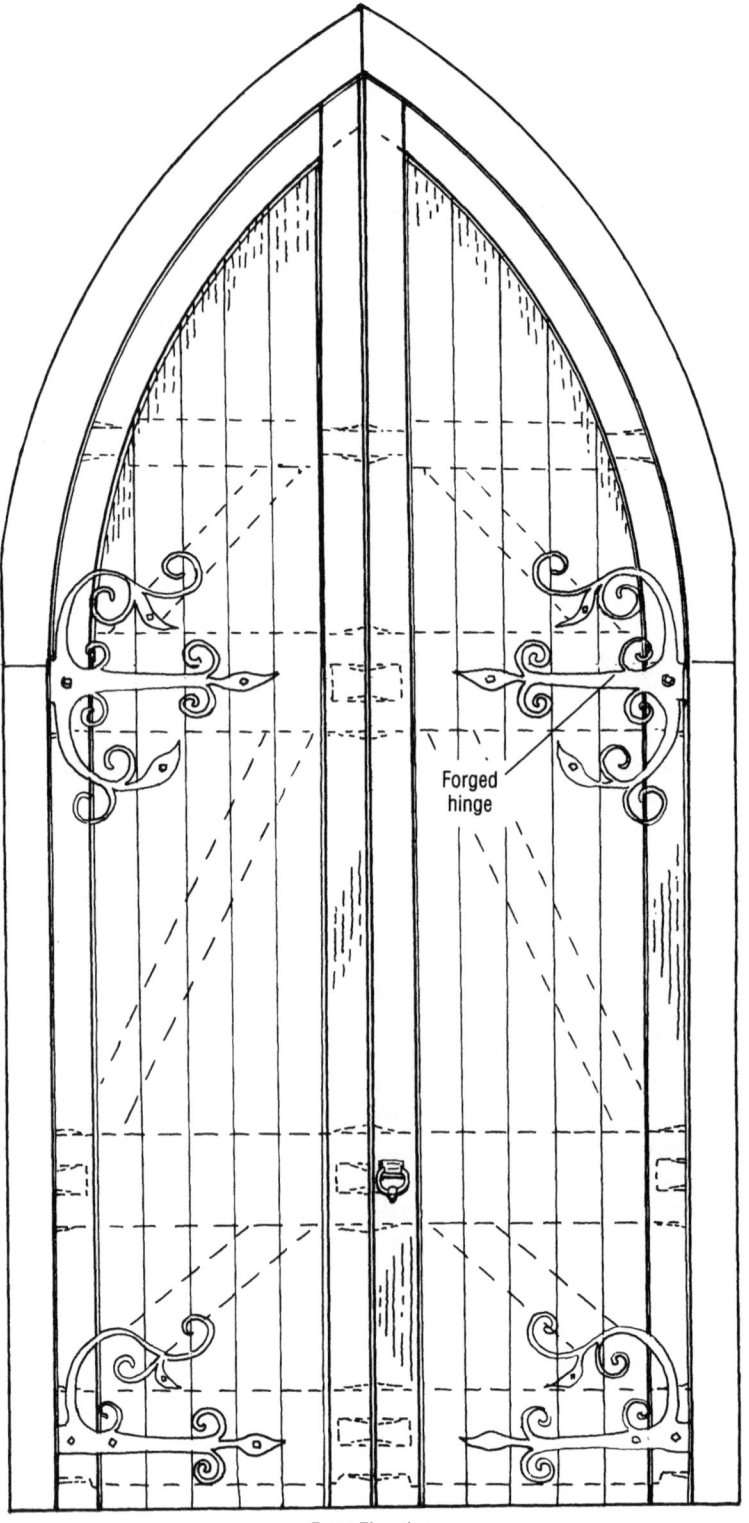

Forged hinge

Front Elevation

28 • *Doors*

England—to use stable-type doors as main entrance doors. The idea is that the top of the door can be hooked open so that you can let some sun in and speak to passersby, and the bottom half can be kept closed to keep dogs and kids in and strangers out. When you want the two halves to function as a single door, you simply latch them together with a couple of tower bolts. The stable door is, in effect, merely a framed, ledged, and braced door that has been cut through the middle ledge, with the top half having a rabbet or rain drip on the bottom edge. The disadvantage with this type of door is that both halves need to be hung so that they open outward, because the top half has to lap over the bottom half.

Framed, ledged, and braced stable door, showing a front or outside elevation and a side section.
Height: *6 feet 6 inches.*
Width: *3 feet.*
Thickness: *2 inches.*
If you are looking for a door for your workshop, a stable, or even a cottage, this is a good choice. Note that the design is such that the top of the door needs to open outward, with the bottom half to follow. The pierced vent feature in the top half of the door is an option that is good for a stable but not so wonderful for a cottage door. You could modify the vent and fit it out with a sliding shutter with folk-art piercings—hearts, tulips, or some such.

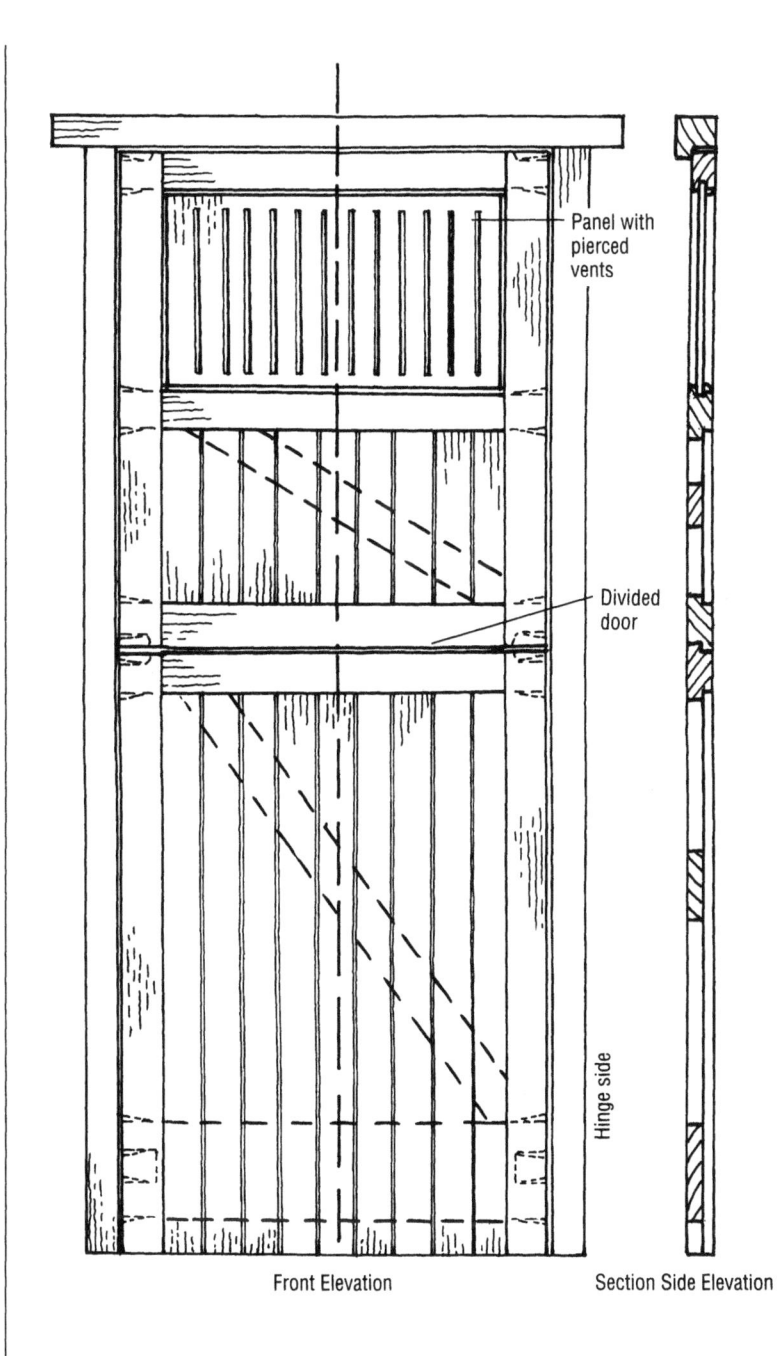

Front Elevation   Section Side Elevation

Generic Door Types • 29

FRAMED AND CROSS-BRACED DOORS

A door of this type has a frame as already described—with stiles and rails—with the addition of vertical timbers called muntins or of cross braces.

Framed, ledged, and cross-braced door, showing a back or inside-house elevation, a section plan, and a side section.
Height: *6 feet 4 inches.*
Width: *2 feet 10 inches.*
Thickness: *2 inches.*
Battens: *4¼ inches wide and ⅝ inch thick.*
This makes a very decorative cottage door for interior or exterior use. The door as illustrated is fitted out with a drip at the bottom, for exterior use. Note that all the members that make up the frame are chamfered so that each little enclosure within the frame has a strong decorative emphasis. Although the door is structurally sound, with good strong ledges and braces, the stiles and head may be a bit skimpy—too narrow for the width of the ledges and braces. If so, the stiles and head can be beefed up. One such door was painted cream, with the chamfers picked out in red. It looked stunning—very William Morris and heraldic.

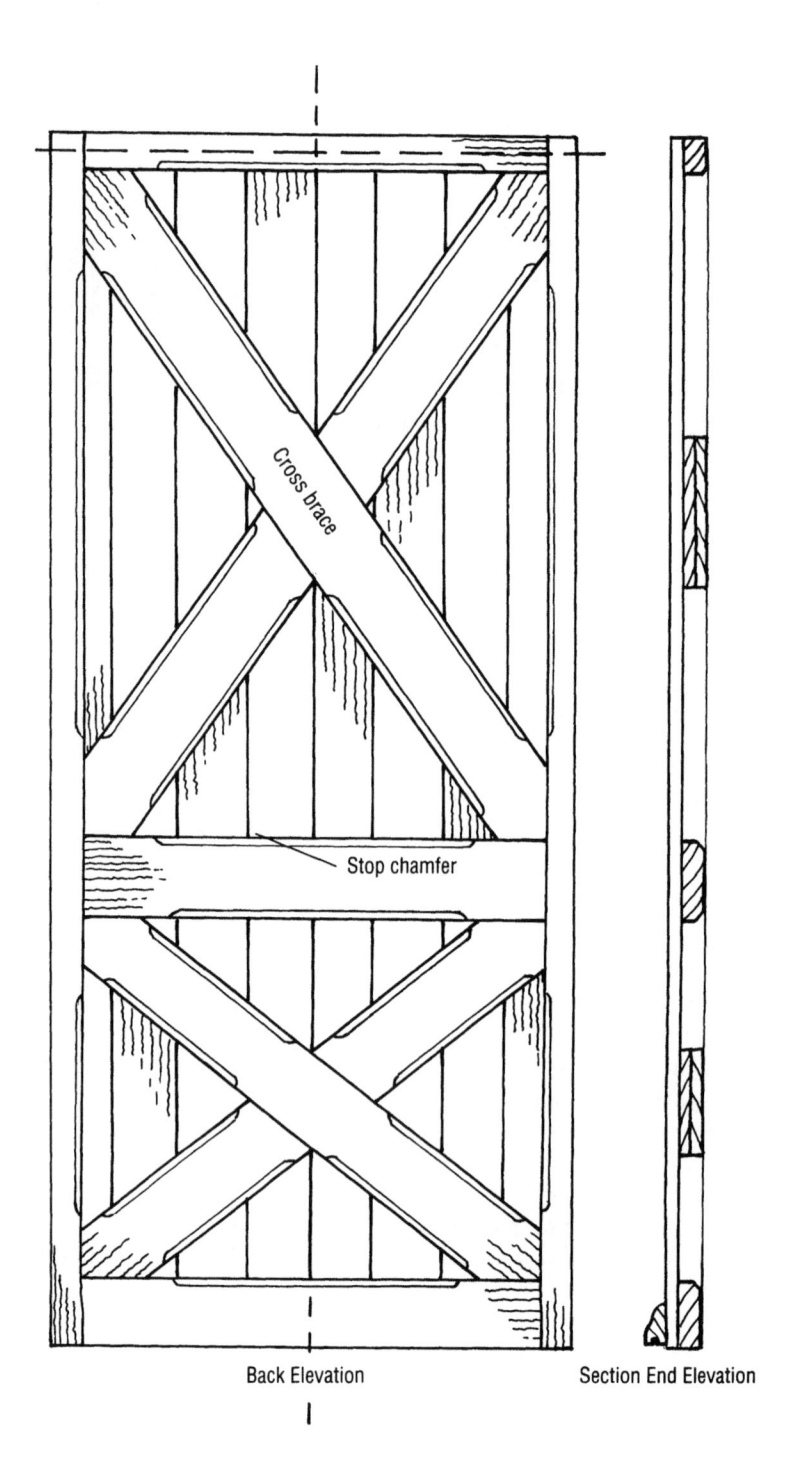

Back Elevation

Section End Elevation

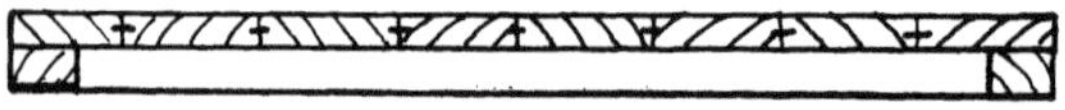

Section Plan

FRAMED AND CROSS-BATTENED DOORS

The frame of this door is much the same as already described, the only difference being that the battens are arranged diagonally. The option is both highly decorative and strong, with each of the diagonal battens becoming an additional brace. If you are looking for a door that is both strong and decorative—one with a Gothic feel—this is a good one.

Framed and cross-battened door, showing a back or inside elevation, a section plan, and a side section.
Height: *7 feet.*
Width: *3 feet 6 inches.*
Thickness: *2 inches.*
Battens: *5 inches wide and 1 inch thick. This is a large, substantial exterior door—perfect for a prestigious town house or maybe for a barn-type conversion. The frame is rabbeted so that the battens are flush with the frame, and the frame is stiffened up with muntins on the inside. The battens are fitted with loose tongues and arranged diagonally, so that each batten becomes a strengthening brace. There needs to be a gap or easement of about 1/32 inch between each of the battens to allow for expansion.*

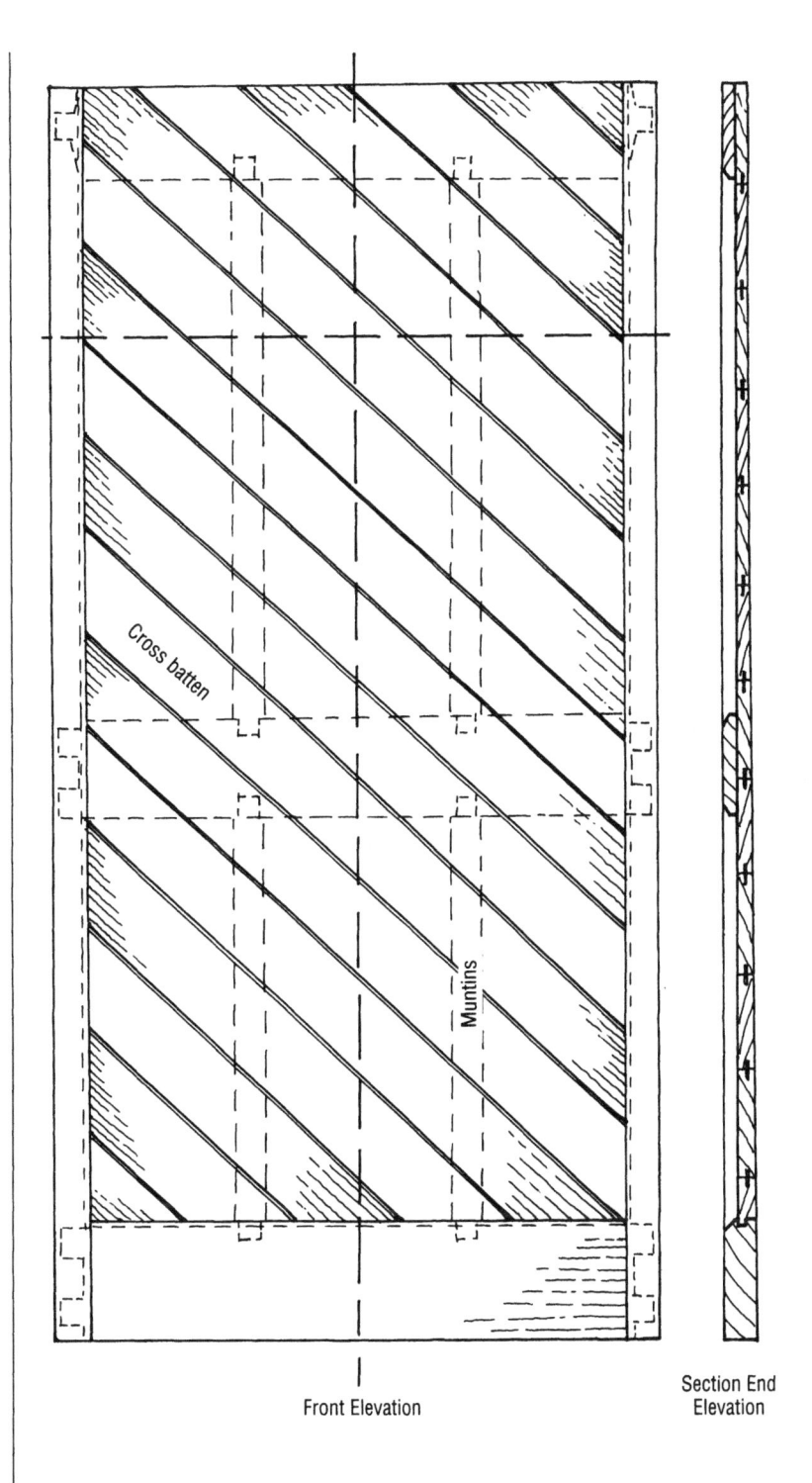

Front Elevation

Section End Elevation

Section Plan

Generic Door Types • 31

SIMPLE FRAMED AND PANELED DOORS

Whereas the framed and ledged door has vertical battens fixed to one side of the frame, the simple framed and paneled door has panels that are fitted into grooves or rabbets worked into the thickness of the framing. The procedures for making framed and paneled doors are very different from those used for building framed and ledged doors. Whereas the ledged door is first framed and then battened—two distinct operations—with the paneled door, the frame and panels need to be worked at the same

Basic framed and paneled door, showing a front elevation, a section plan, and a side section.
Height: *6 feet 6 inches.*
Width: *2 feet 6 inches.*
Thickness: *1½ inches.*
This is the simplest of the framed and paneled doors. The stiles and ledges are flat-faced and unadorned, the panels are flat and rabbeted straight into the frame without the benefit of moldings, and all the specifications are minimal. The top rail is 4 inches wide, the bottom rail and the lock rail are 9 inches wide, and the stiles are 3½ inches wide. When speculative builders at the turn of the century were throwing up the worst sort of dwellings across Britain and the United States, this was the door they used. This type of door is depressingly stark and unpleasant, especially when made of pine. It's not so bad when it's made of pitch pine or cherry.

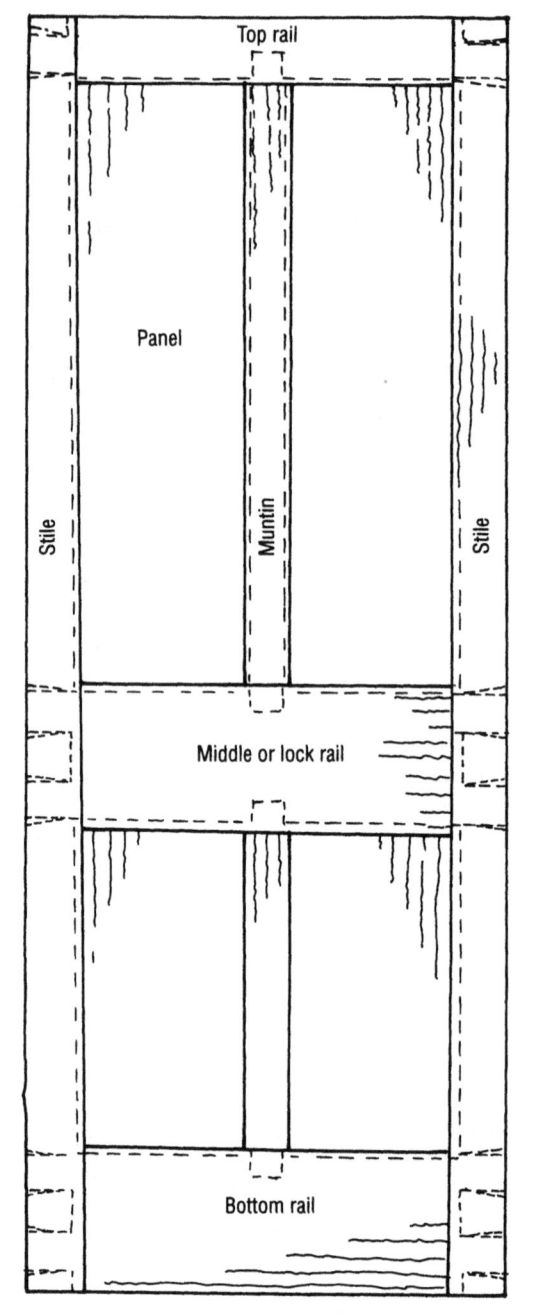

Elevation of Paneled Door

Section Plan

Section End Elevation

32 • Doors

time. This makes for a more complicated procedure, requiring much more planning at the design stage. The basic panel door has a minimal cruciform frame and four panels that are usually one-third the thickness of the frame. The panels fit into grooves that are worked in the middle of the frame.

There are several basic design options:

• The panels can be plain on both sides.

• The panels can be plain on both sides, with the framing stop-chamfered on the better face.

• The panels can be plain on one side, with a modest molding nailed on the better side and fitted within the frame.

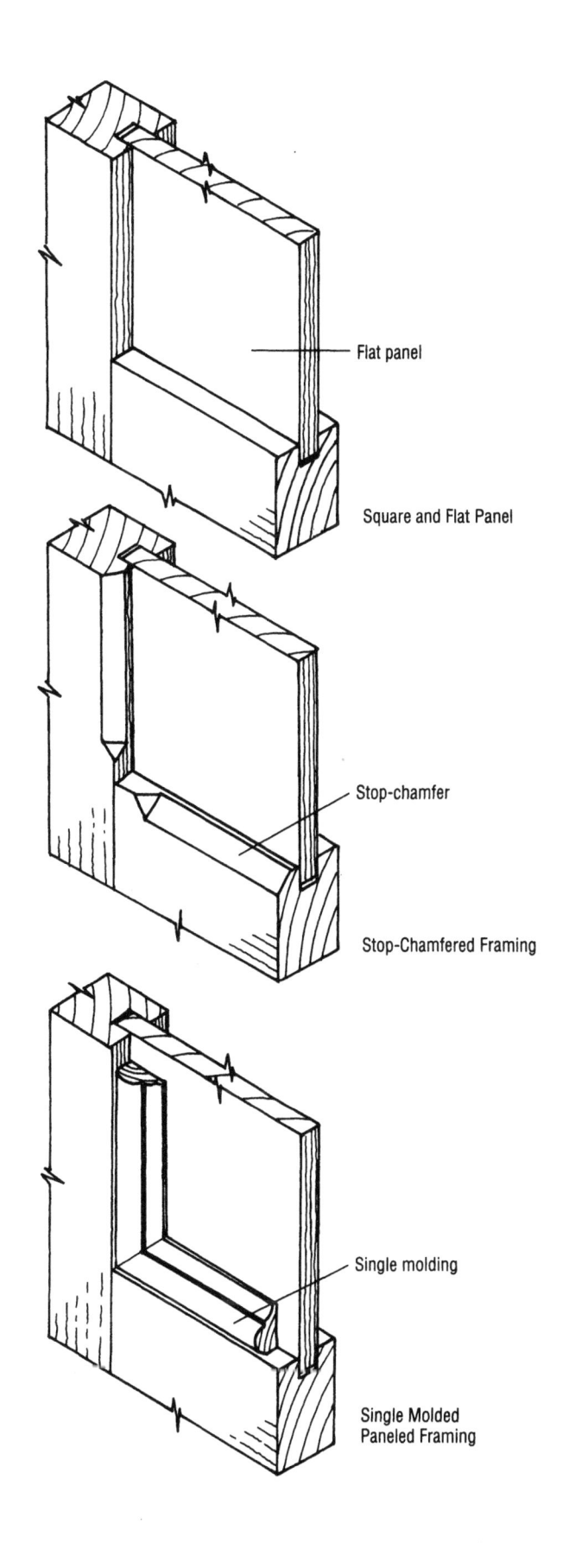

Frame and panel details.
Top: *Square and flat frame and panel—the simplest of the framed and paneled options.*
Middle: *Stop-chamfered frame and flat panel. Apart from the stop-chamfered details on the frame, this door is identical to the square and flat design. The chamfering makes all the difference—it gives the door a bit of character.*
Bottom: *Square and flat framed and paneled door with a modest single molding on one side. The molding is usually fixed to the on-show side—for example, on the side of the linen closet door that faces the hallway.*

Square and Flat Panel — Flat panel

Stop-Chamfered Framing — Stop-chamfer

Single Molded Paneled Framing — Single molding

Generic Door Types • 33

- The panel can be plain on one side, with generous bolection molding nailed to the better side so that part of the molding laps over the frame.
- The panel can be plain on one side, with the panel being a thicker section and rabbeted into the frame and a beading running flush around the edges.

Frame and panel details.
Top: *A framed and paneled door with a generous bolection molding. Once again, the molding is used on the show side of the door. When both faces of the door need to be showy, bolection moldings are used on both sides.*
Bottom: *A frame and panel with bead-flush beading. With the panel being two-thirds the thickness of the framing, the beading is worked from the thickness of the panel on the vertical edge and planted in a groove on the horizontal end-of-grain edge. Some woodworkers run an extra-wide rabbet on the panel edges so that there is room between the inside edge of the frame and the raised part of the panel for a planted molding.*

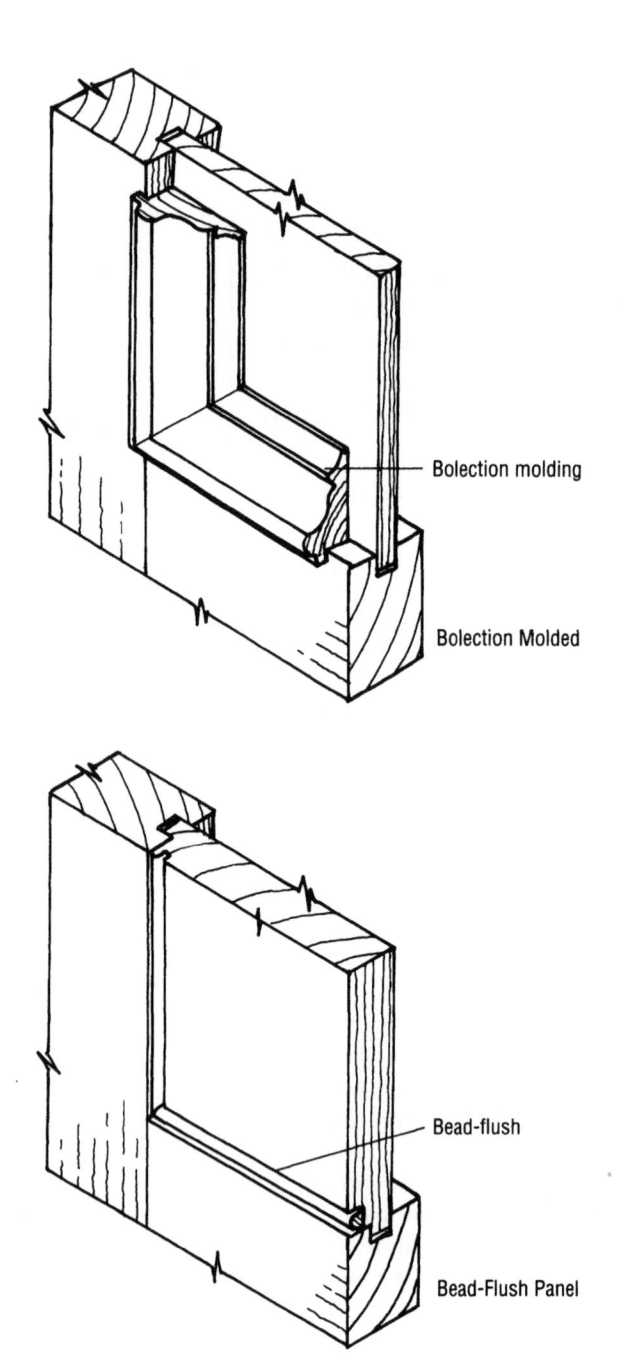

SPECIAL FRAMED AND PANELED DOORS

The term *special* is used in this book to define all framed and paneled doors that, because of their size, shape, or specifications, go beyond the simple.

Four-Panel Door

The special four-panel door is much the same as the simple four-panel door already described; the only real difference is that all the specifications are considerably more generous. There is a solid-section bolection molding running through the thickness of the door, and the section is tongued into the frame rather than being nailed on. Although this detail might not seem special, it makes for a door that is superior in both strength and design. A door of this type is a joy to the eye and will last a lifetime.

Methods of fixing moldings into the framing.
Top: *On top-quality work, both the frame and the molding are grooved, so that when the door is put together, the panel and the molding are held in place without the need for nails or screws. The resultant molding is known as double bolection through-frame molding. Although this is arguably the best way of fitting molding, the panels and the moldings must be cut to perfection.*
Bottom: *A good option for fixing a generous bolection molding to one side of the door and a plain molding to the other is to have both moldings set in grooves as shown. If the whole thing is going to come together for a good fit, all the grooves and rabbets must be worked with great care.*

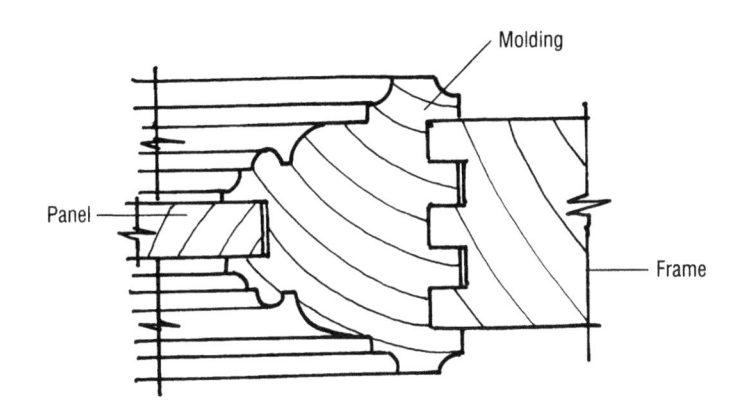

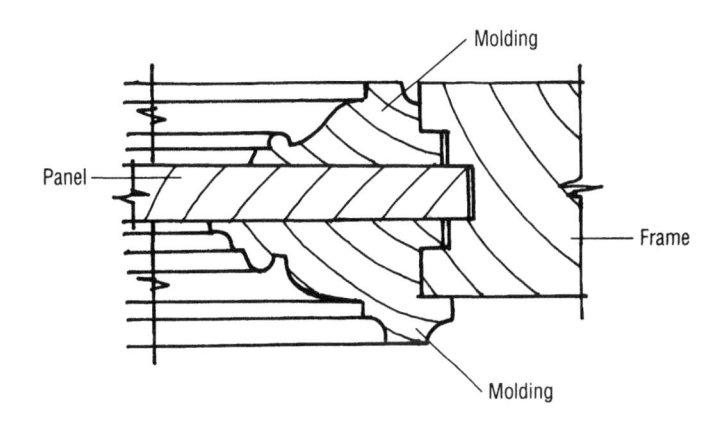

Six-Panel Door

The six-panel door is identical in all respects to the four-panel door, except for the number of panels. When you need a larger door—that is, one that is a little bit wider and a little bit taller—the whole structure needs to be beefed up. This means that the sections have to be thicker and another rail has to be added. This usually involves dividing up the top panel and having an extra rail at about head height. From bottom to top, the component sequence is now bottom rail, bottom panel, lock rail, middle panel, frieze rail, frieze panel, and top rail. The additional panels don't affect the making of the door—apart from requiring more time, effort, and wood—but they allow the door to be much taller without affecting the balance of the design.

Framed six-panel interior door, showing a front elevation, a section plan, and a side section.
Height: *6 feet 9 inches.*
Width: *2 feet 9 inches.*
Thickness: *1¾ inches.*
The stiles and the top rail are 4 inches wide, the frieze rail is 3½ inches wide, the lock rail and the bottom rail are 10 inches wide, and the center of the lock rail is set 2 feet 9 inches up from the bottom of the door. Although this door is only 1 inch higher than a standard four-panel door, it's clear that the addition of the frieze rail breaks up the height and improves the overall proportions. This particular door has plain moldings on both sides. If you want an elegant interior door and are considering using decorative woods with an oiled finish—something like cherry or cedar—the six-panel door is a good choice.

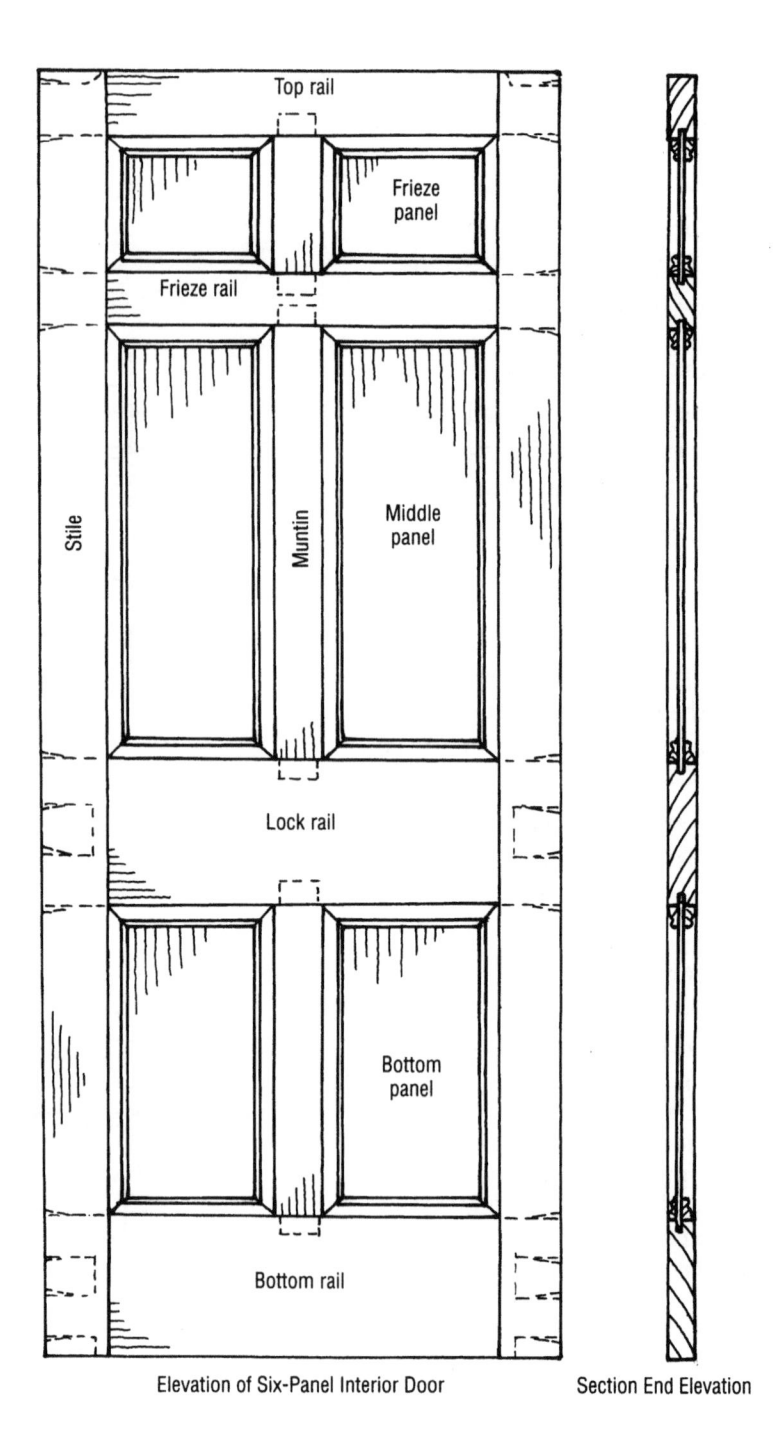

Elevation of Six-Panel Interior Door

Section End Elevation

Section Plan

36 • Doors

DOUBLE-MARGIN DOORS

When an opening is too wide for a single-width door (at about 3 feet 11 inches) yet too narrow for a pair of double doors—because one door is too narrow to allow easy passage—the traditional answer is to make two completely separate narrow doors, fix them together with wedges, and hang them as a single door. Although the result is a single door hinged at one side, the overall visual effect is one of a pair of attractive narrow doors.

DOUBLE DOORS

When a doorway is too big for a single-width door or even a double-margin door, the answer is to have a pair of double doors—two true doors that are hinged one at each side of the opening to meet at the middle. Although this option seems pretty straightforward, the design proviso is that each door must be wide enough to allow easy passage.

Double-margin doors, showing a front elevation and two section plan details. The whole idea of a double-margin door is to make a single door that looks like two doors. The problem is not that it's impossible to create a wide six-panel door—because it's pretty obvious that you can—but rather that doors that grow in width without growing in height tend to look squat and generally out of proportion. Note that although the folding wedges are great for holding the two narrow doors together, a flat metal bar needs to be let in at the top and bottom to help prevent the door from flexing across the width.

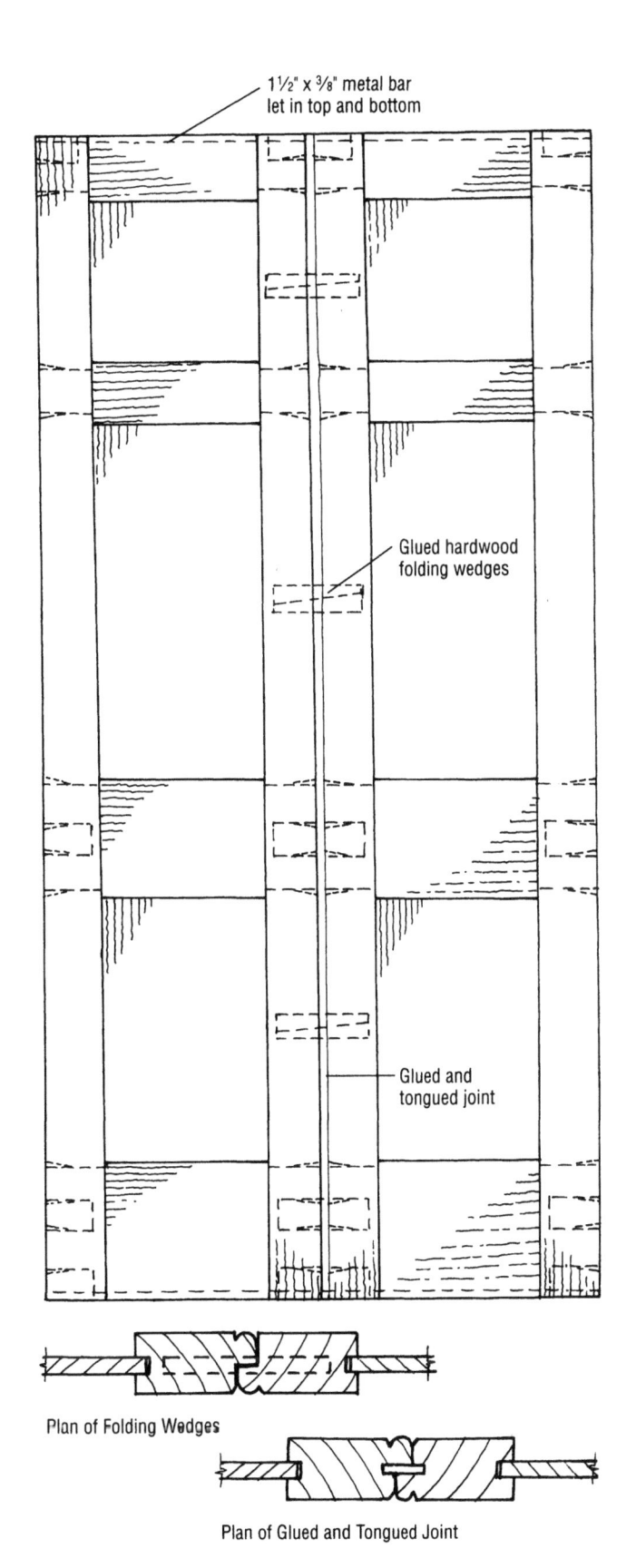

1½" x ⅜" metal bar let in top and bottom

Glued hardwood folding wedges

Glued and tongued joint

Plan of Folding Wedges

Plan of Glued and Tongued Joint

Generic Door Types • 37

For example, if you had double doors in a narrow opening, with each door being less than about 26 inches wide, the whole thing would be an inconvenience, because you would always have to open both doors to go through.

SASH DOORS

Sash doors traditionally have glass panels in the upper half. Such doors are usually used when extra light is needed or it is important to be able to see through the door when it is closed. Doors of this character can be in the form of simple four-panel doors, with the top two panels made of glass rather than wood, or they can be grand doors with the glass stained and ornate. Double doors and double-margin doors can also be made in the sash form. When a sash door is used as an exterior door, weather bars and drips are usually needed to shed water.

Four-panel door with leaded light upper panels, showing a best face elevation and two detailed sections:
Height: *7 feet.*
Width: *3 feet.*
Thickness: *1¾ inches.*
Panel thickness: *1 inch.*
This good-quality sash door conversion—meaning a four-panel door that has been made into a sash door—has bolection moldings at the better face and bead-butt moldings on the side edges of the second-face panels. If you want to turn one of your existing four-panel doors into a relatively well-proportioned sash door with glazed leaded lights, this door is a good option.

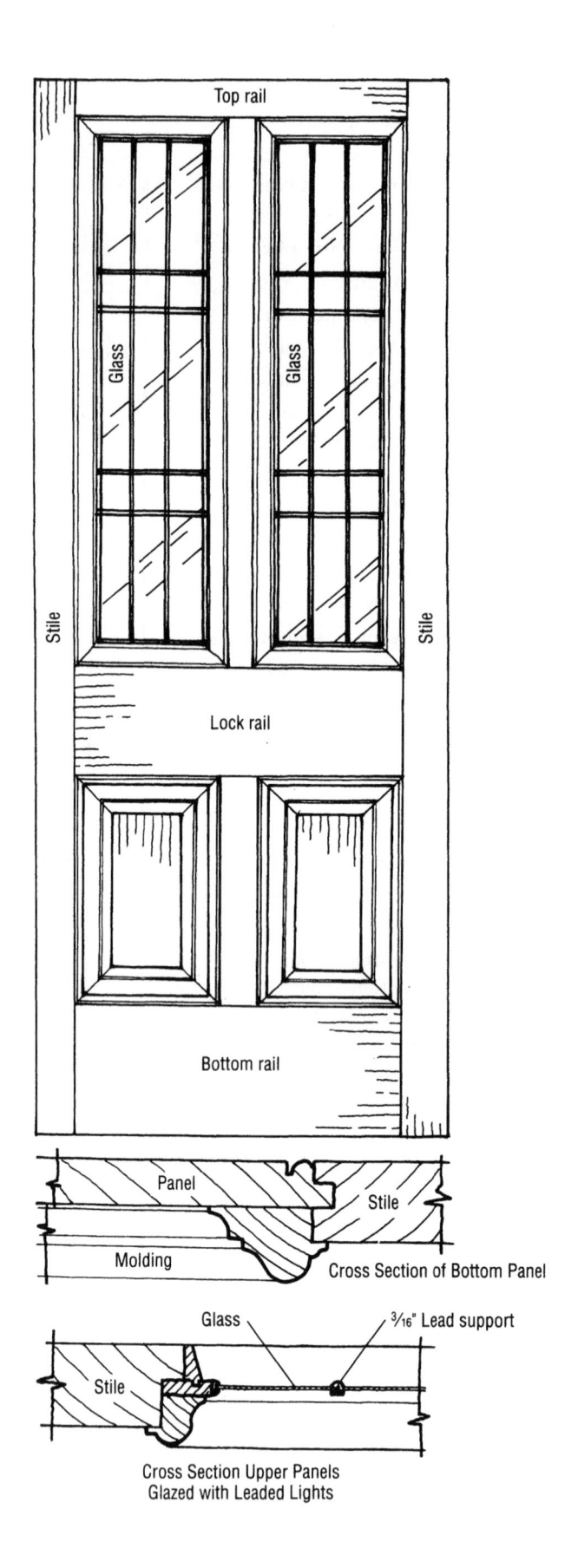

Sash door with diminished stiles, showing a front elevation, section plan, and side section, with glazing bar details.
Height: 6 feet 5 inches.
Width: 2 feet 8 inches.
Thickness: 2¼ inches.
This is your first choice when you need a fancy door with maximum viewing or lighting in the upper half. Note that the shape of the reduced stile—sometimes called a gunstock stile—looks very much like a rifle stock in side view. The top of the stiles are reduced to the same width as the top rail, with the beaded glazing bars stub-tenoned into the stiles and the rails. The details show how the glazing bars are halved and mitered into each other.

DIMINISHED STILE DOORS

Sometimes called gunstock stile doors, this type of door gets its name from the fact that the stile decreases in width from the top of the middle or lock rail through to the top rail. Diminished stile doors are sometimes referred to as sash doors, for the simple reason that most doors of this type are glazed.

INTERIOR AND EXTERIOR DOORS

Although there is nothing to keep you from building the grandest exterior door and hanging it inside the house, there is every reason not to use an interior door in an exterior opening. The reason is that exterior doors are built to higher specifications, with choice wood, heavier sections, and bigger hinges and locks; they are fitted with various drips and rabbets to throw off the water and hung so that they open inward.

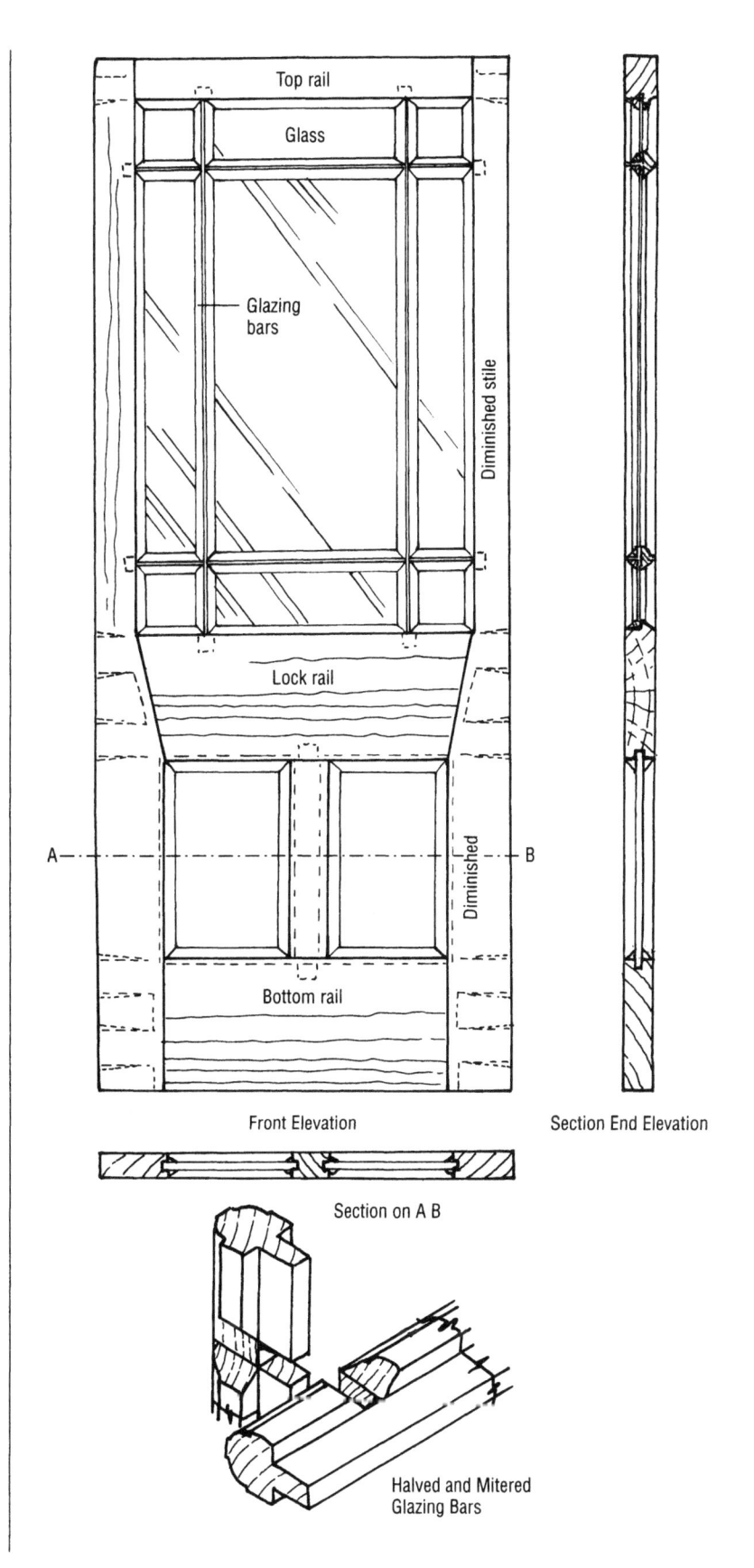

Generic Door Types • 39

Making a Ledged Door

Although the ledged door is certainly the simplest form of door—swift and easy to build with a minimum of tools and expertise—it can be built to a very high standard. The ledged door has a long and noble tradition that goes way back. If you build it well from wide oak boards with clinched nails, good-quality cross-garnet hinges, and maybe a nice old hand-forged Suffolk or Norfolk thumb latch (relatively easy to obtain from one of the many architectural salvage companies), it will easily outlast you. Start the project by sitting down and considering your needs. Ask yourself the following:

- Is the door you have in mind appropriate for its purpose (its location and its structure)?
- Do you want to make the cheapest door you can (from off-the-shelf ¾-inch, 4-inch stuff), or do you want to make a good solid oak door with 9-inch-wide, 1½-inch-thick boards?
- Do you want a fancy traditional door with the ledges at different widths (say the top ledge at 6 inches, the middle lock ledge at about 9 inches, and the bottom ledge at 8 inches), or do you want standard 6-inch ledges throughout?
- Are you trying to make an authentic country door with authentic forged iron fittings, or are you simply after an inexpensive look-alike?
- Are you going to use traditional hand tools or power tools?
- Do you have the correct tools?

Carefully consider all the pros and cons of the project—design, costs, tools, and the like—until you have a clear understanding of all the implications of making the ledged door of your choice. The illustration on page 41 shows a high-quality oak door that will last a lifetime.

TOOLS
- Two saws—a ripsaw and a crosscut saw.
- Three bench planes—a smoothing, a jointer, and a block.
- A pair of wooden tongue-and-groove planes to suit the thickness of the wood, or a Stanley 45 combination plane.
- A hammer and a nail set or punch.
- A pair of sash clamps.
- A bench large enough to take the door.

CUTTING LIST AND FIXTURES
- Three ledge boards about 1½ inches thick and 6 to 9 inches wide at a length to suit the width of the door.
- Six battens or boards at 1 inch thick and 5 to 6 inches wide at a length to suit the height of the door. Make sure that there is plenty of width so the boards can be cut back.
- A pair of hinges of the strap or cross-garnet type.
- A number of screws or bolts to fit the hinges.
- A quantity of rose-head clinch nails about 2½ to 2¾ inches long, variously described as wrot, wrought, clasp, or even clinching nails. *Note:* Though you must use wrought iron nails—because the fibrous nature of the iron allows the point of

the nail to curl over—be ready for the weird and wonderful traditional names that these nails are sold under.

DOOR-MAKING STEPS

1. Having sawed all the wood to size—leaving a generous allowance for cutting waste—plane the six boards to smooth both faces and to joint, square, and true both edges.

2. Set the boards side by side and number them in pencil from left to right so that you can reset them in the same order.

3. Clamp the boards together and measure the width. Make allowances for the fact that the tongues will be lost in the grooves; decide how much needs to be cut away, and then reduce the width of boards 2, 3, 4, and 5 accordingly.

4. Use a pair of matched tongue-and-groove hand planes, or a combination plane, to cut the tongues and grooves on the appropriate edges of the boards. Be mindful that the outside edges of both outer boards need to be free of tongues and grooves.

> **SPECIAL TIP**
> If you want a more decorative door, try one of the fancy bead and rabbet options. There are many exciting permutations.

5. Use a plane to chamfer all the mating edges so that the boards meet to make a V-section groove on both the front and back faces.

6. Clamp the six boards together and measure for width. If you've done it right, the outer boards will be a good deal wider than the inner boards, and the total width will be slightly wider than the doorway.

7. Take the three ledge boards, cut them to length so that they are about an inch shorter than the width of the door as it sits between the jambs (so that they clear the jambs), and then plane all faces, edges, and ends to a smooth finish.

8. Use a plane to bevel and chamfer all the best-face edges so that the chamfer runs around all four sides of the best face.

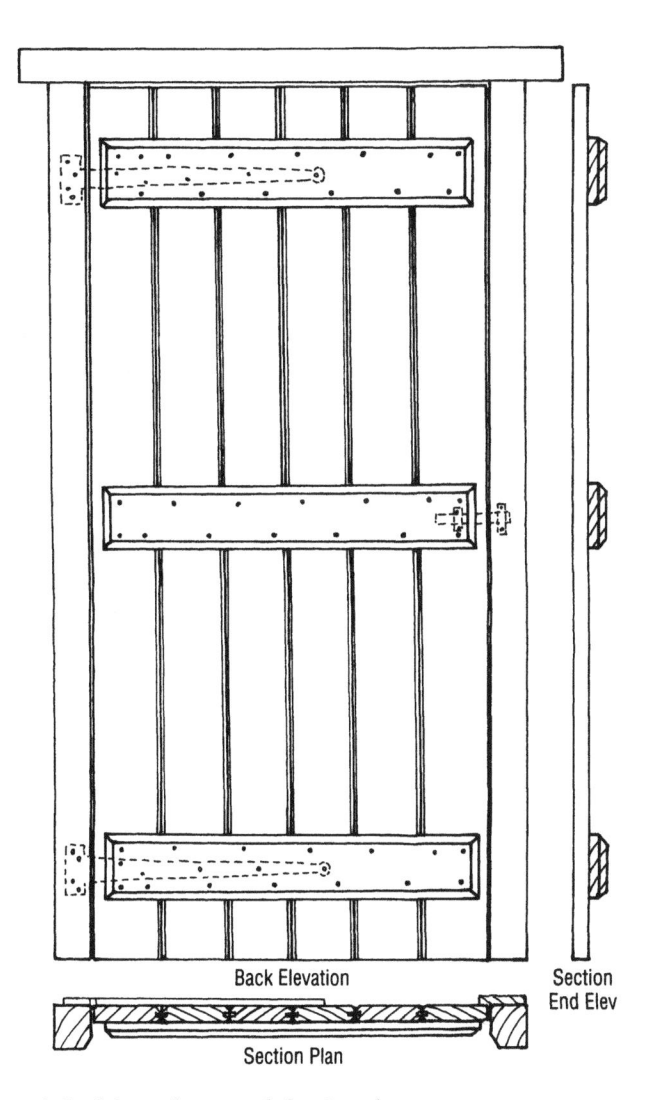

A ledged door, showing a ledge-face elevation, a section plan, and a side section.
Height and width: *To suit.*
Ledges: *6 to 9 inches wide and 1½ inches thick.*
Battens: *5 to 6 inches wide and 1 inch thick.*

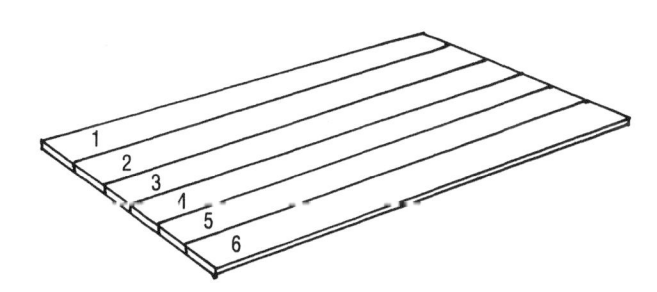

Set the boards together, arrange them for best effect, and number them.

Making a Ledged Door • 41

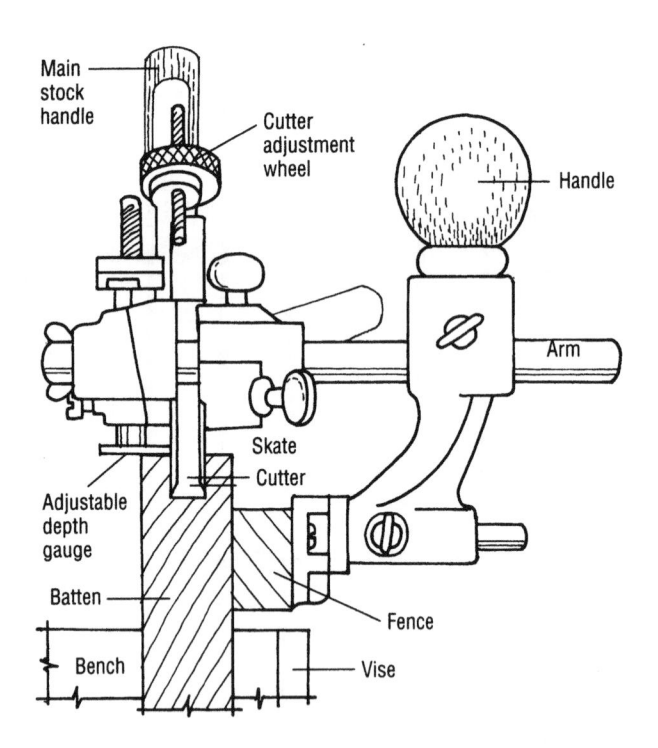

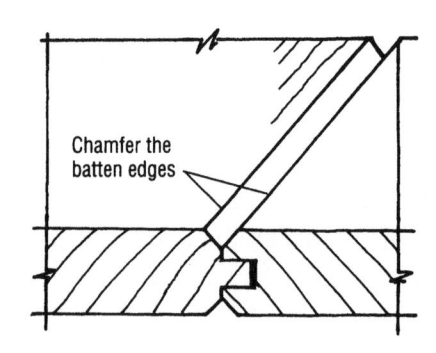

The simplest form of decoration is to plane off the edges to create a chamfer. It's easy to do and can be done in moments with a block plane.

If you enjoy making doors slowly and quietly—with a minimum of fuss, noise, and wood dust—you need to get an old Stanley 45 combination plane. This is the perfect tool for cutting grooves, tongues, rabbets, and beads. You clamp the wood in the vise, set the fence so that the cutter is centered on the wood, and then remove the waste as a series of thin shavings.

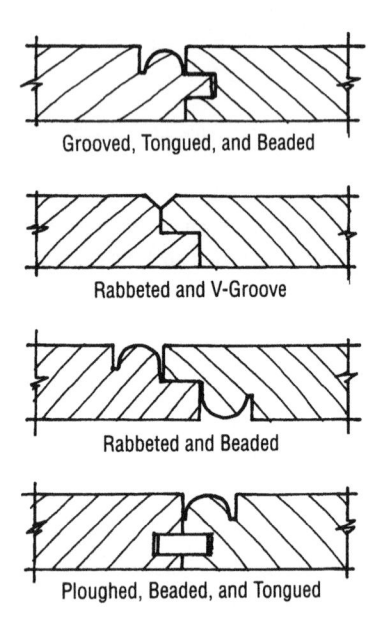

9. Set all six boards together, slide plastic strips that are about 1/16 inch thick between mating boards (on the best face), and then lightly clamp.

SPECIAL TIP
The idea behind the plastic spacers is to allow the wood to expand. This is important when making exterior doors. Although you could use strips of cardboard, they might stick to boards that have been freshly painted just before clamping.

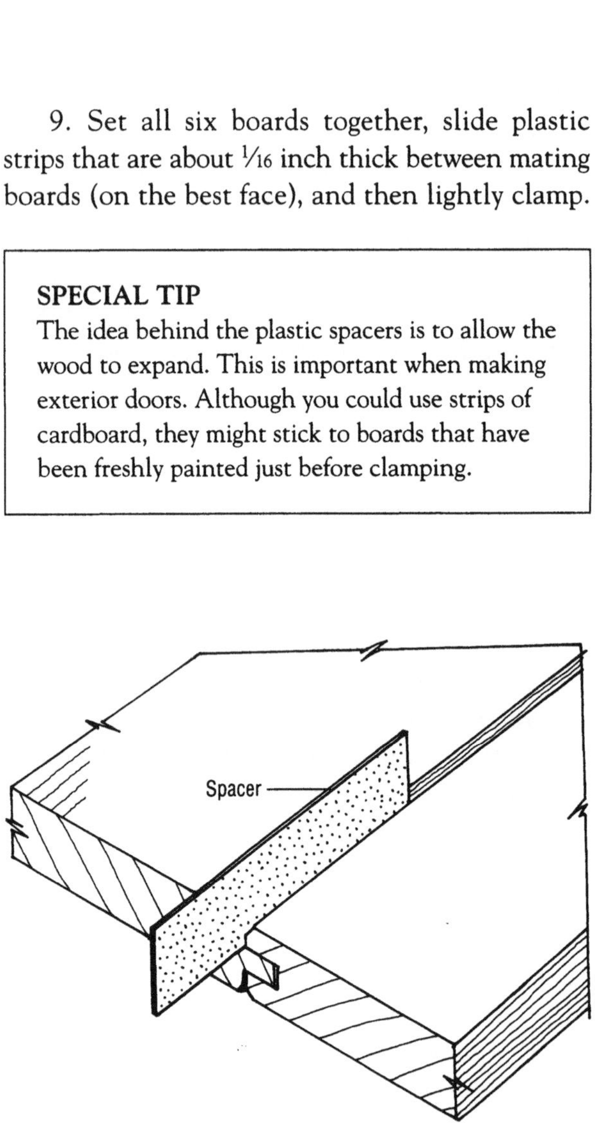

You don't have to settle for a simple tongue and groove. You can have a decorative bead to the side, or you can leave out the tongue and the groove altogether and go for lapped rabbets.

Have plastic spacers top and bottom between mating battens. A total of five spacers across the width creates an expansion easement of 3/8 to 1/2 inch.

42 • Doors

10. Turn the whole works over so that the best face is down on the bench, and then carefully place and align the three ledge boards. Run two short temporary nails partway through the ends of all three ledges so that they are fixed to the outer boards.

> **SPECIAL TIP**
> If you intend to make a number of doors, it's a good idea to make a framing bench. Otherwise, you will have to set a couple of substantial timbers across your bench so that you don't end up with the door nailed to the bench.

11. Mark the position of the ledges through to the outer boards, turn the door over so that the best face is uppermost, and use a square to run the edge-of-ledge guidelines across the width of the door.

12. With the best face uppermost, take a hammer and work along the marked-out ledges (on the best face of the door) and run rose-head nails through the board and on through the ledges.

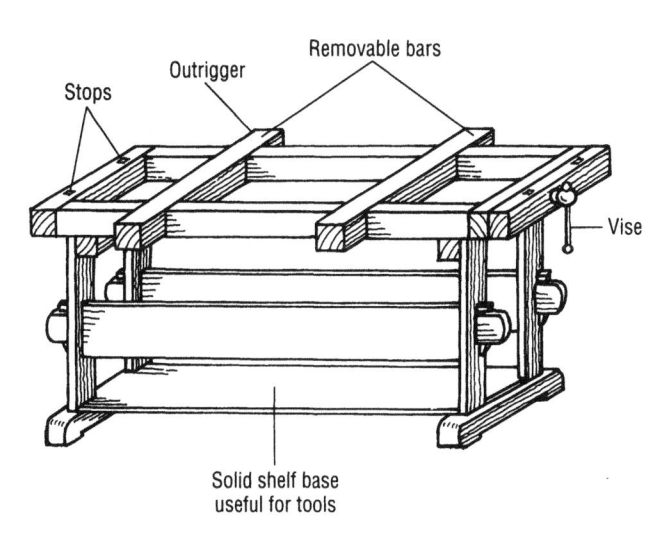

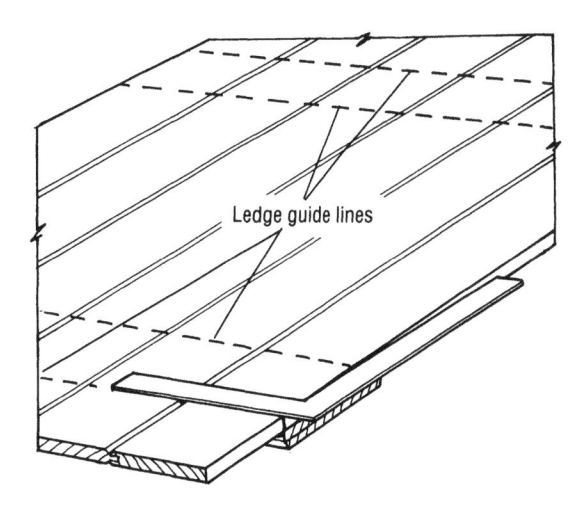

Use a large metal square to run width-of-ledge guidelines across the best face of the door. Note that one arm of the square is tilted so that it rests on the edge of the door.

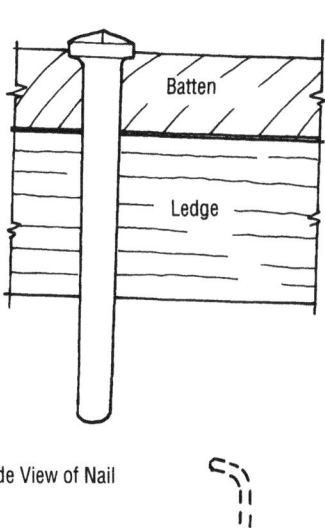

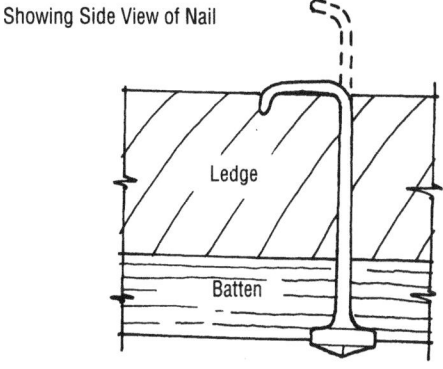

A framing or door-making bench is a pleasure to use and makes the whole procedure much easier. This particular design has a vise at one end, stops at both ends, and two outriggers across the width to support wide doors. The door is set in place and caught between the stops, and then sash clamps are used across the width. The open frame allows easy access to and viewing of the underside.

Top: *Drive the rose-head nail right through the wood.*
Bottom: *Encourage the nail to roll over at the point, and then use the punch to drive the tip of the nail back into the wood. Be careful not to mash and do damage to the wood, and don't try to bury the whole end of the nail. It's much better to leave the bridged and clinched end as a feature.*

Making a Ledged Door • 43

Drive the nails home so that ¾ to 1 inch of the point sticks through.

13. Turn the door over and clinch the nail points over. First bend the points over so that they run at right angles across the run of the grain, and then use a nail set to punch the point over, down, and around into the wood.

14. When the door is all nicely clinched and square, remove the temporary nails, saw the door to length, and plane it to a finished width.

15. Finally, set the cross-garnet hinges in place on the top and bottom ledges, fit them with screws or bolts, screw the door in place, and fit the latch.

AFTERTHOUGHTS AND OPTIONS

- Very old ledged doors have been made from two oak boards about 1½ inches thick, with ledges about 12 inches wide and 2 inches thick and everything clinched with massive hand-forged nails. They can be impressive.
- To a great extent, the character of the door depends on the size and choice of wood. Wide, thick boards make for better proportions.
- There are two schools of thought about clinching nails. One holds that the points should be bent over so that they are in line with the grain—so that the bent-over point is hammered into the grain. The other school holds that the points should be bent over so that they are across the grain and highly visible. The latter technique is preferable because there is less chance that the point will split the wood and a greater chance that the clinching will hold.
- If you intend to use the door in the house, store the wood inside for a few weeks so that it has a chance to adjust to the humidity.
- A good option for making an authentic restoration door is to use floorboards salvaged from an old house. If you're lucky, you'll find floorboards that have the tongues and grooves intact.
- If you intend to make lots of doors, get a pair of matched planes that are dedicated to cutting tongues and grooves or a combination plane of the Stanley 45 type that is fitted with tongue-and-groove cutters.
- A good traditional method of preserving an exterior door from the ravages of weather is to paint all mating faces and edges before clamping—the backs of the ledges, inside the tongues and grooves, and so on.

Making a Four-Panel Door

Traditionally, good-quality interior and exterior doors are usually paneled. A typical traditional door consists of a grooved framing designed to take panels of thinner wood, with the panels being held in place with moldings. The number of panels and the size and complexity of the molding depend on the size of the door and the whims and fancies of the designer and the client. The run-of-the-mill panel door—the one found in most ordinary three-bedroom homes in the nineteenth century through the 1940s—is the four-panel door.

The common four-panel door ranges in size from 6 feet 6 inches high by 2 feet 6 inches wide to 6 feet 8 inches high by 2 feet 8 inches wide. The frame varies in thickness from 1¼ to 2¼ inches, and there are plain or fancy moldings to suit. The following project describes how to make a door 6 feet 8 inches high by 2 feet 8 inches wide with a 1½-inch-thick frame, with plain molding on one side and fancy bolection molding on the other.

As with all the projects, you need to start out by carefully considering your needs. Before putting tools to wood, you must spend time pondering the following:

- Do you want a door that is slightly bigger or smaller than the one described? If so, how will it change the project—the wood sizes and the tools used? Will it make any real difference to the procedures?
- Do you want a paneled door that is taller than average? If so, all you do is increase the height and split the top panel with a frieze rail so that you have six panels.
- What type of wood are you going to use? Do you want the door plain or painted? How will this affect your choice of wood?
- Do you have all the necessary tools?

Spend time with a pencil and paper until you know precisely what you're doing. Make sketches and notes, and draw up a cutting list.

SPECIAL TIP
If you want to make two or more identical doors, mark up the components in batches—all the stiles, all the top rails, and so on. This speeds up production.

TOOLS

- Four saws—a ripsaw or half ripsaw for ripping down and for cutting the tenons, a crosscut saw for sizing, a small fine-bladed backsaw for mitering the molding, and a coping saw for clearing the waste between the paired tenons.
- Five planes—a smoothing, a jointer, a block, a rabbet or shoulder, and a plough or combination plane for cutting the grooves.
- A pencil, rule, mortise gauge, knife, and square for marking.

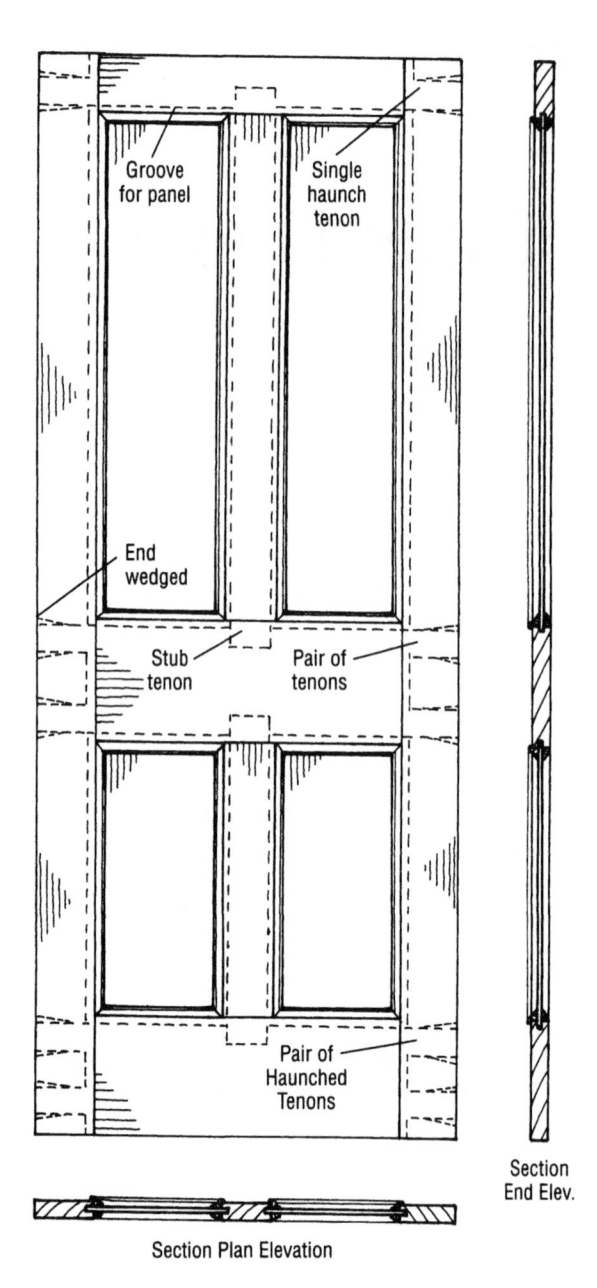

A classic four-panel framed and paneled door, showing a second face elevation, section plan, and side section. The height of the top edge of the middle rail is usually set about 3 feet 2 inches from the ground so that the door handle comfortably meets the hand.

- A mortise chisel approximately one-third the width of the frame thickness.
- A mallet and two hammers—a claw and a pin hammer.
- A pair of sash clamps.

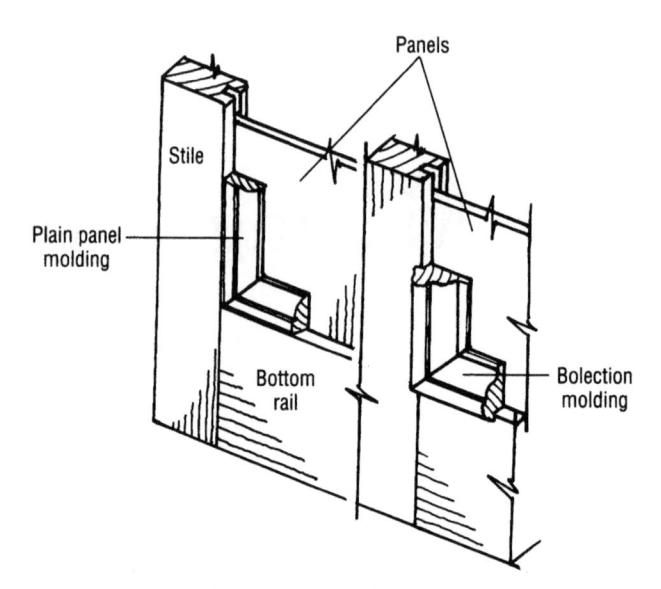

A cross section through the door showing the two faces. Note how the two moldings are fixed—the bolection so that it laps over the edge of the frame, and the plain so that it sits within the frame. The bolection molding is usually the first choice for exterior doors, because the rabbeted edge helps throw off water.

CUTTING LIST AND FIXTURES (FOR ONE DOOR)

- Two stiles 6 feet 10 inches long, 5 inches wide, and 1½ inches thick. It's best to use a straight-grained wood such as redwood or American oak.
- A head rail 2 feet 8½ inches long, 5 inches wide, and 1½ inches thick of the same type of wood as the stiles.
- Two rails—one for the lock and one for the bottom—2 feet 8½ inches long, 10 inches wide, and 1½ inches thick.
- Two muntins—one 2 feet long and the other 3 feet 3 inches long, and both 5 inches wide and 1½ inches thick.
- Two top panels 3 feet long, 10 inches wide, and ½ inch thick.
- Two bottom panels 1 foot 10 inches long, 10 inches wide, and ½ inch thick.
- A 54-foot length of good-quality bolection molding to suit your door—for the best side.
- A 54-foot length of plain panel molding for the second side.

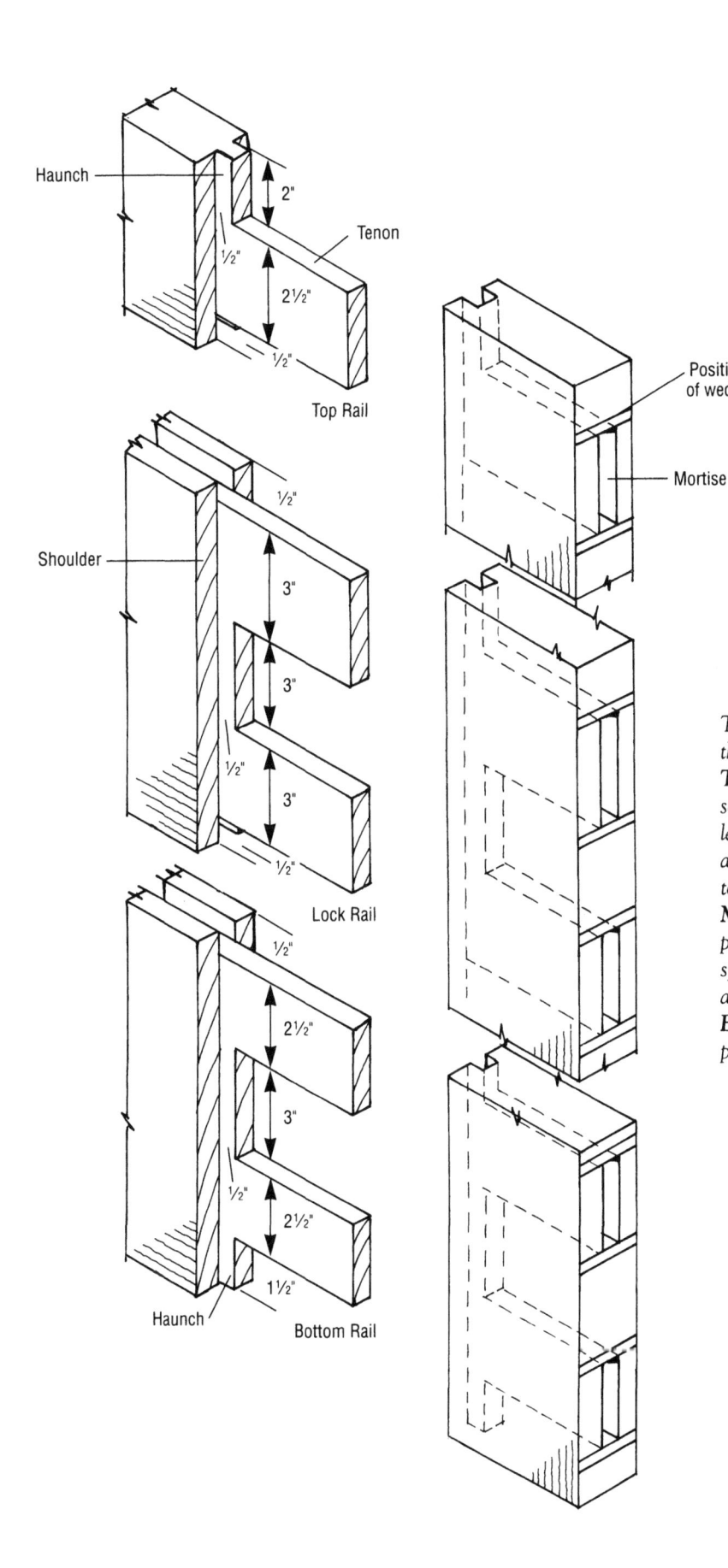

The framing joints between the rails and the stile.

Top: *The top rail is haunched at the top, so that the top end of the stile is more or less intact. Note the thickness at the top and bottom of the mortise, which will need to be cut back.*

Middle: *The middle rail is jointed with a pair of single tenons. Note the "third" spacing across the thickness of the stile and across the width of the rail.*

Bottom: *The bottom rail is jointed with a pair of single haunched tenons.*

Making a Four-Panel Door • 47

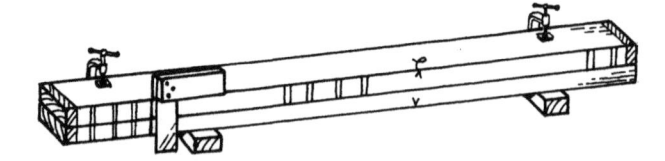

Use the square to run the lines over both pieces of wood.

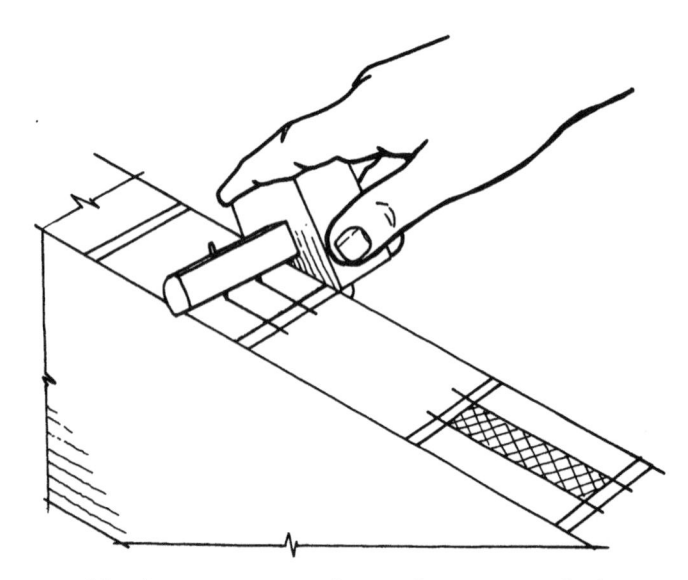

Use the mortise gauge to lay out the mortises on the face side of the stiles. It might help to shade in the areas that need to be removed.

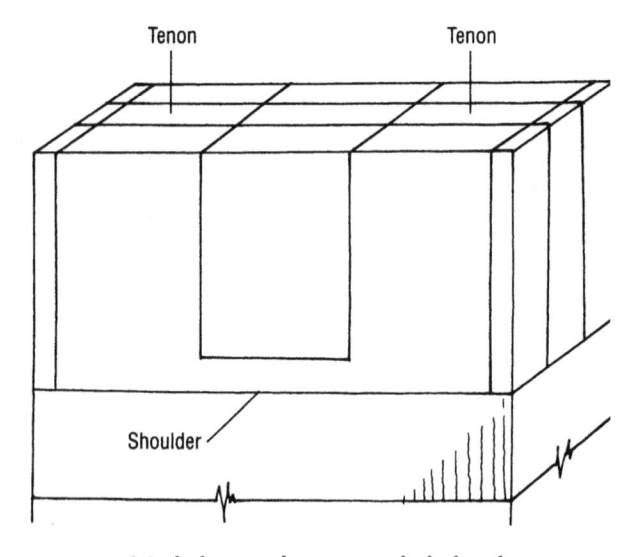

Mark the paired tenons on the lock rail.

DOOR-MAKING STEPS
Marking

1. After drawing up the design and sorting the wood to best advantage—to avoid small flaws, loose knots, splits, and the like—cut the wood to length and plane all the stiles and rails to a good smooth finish. Mark the best-face side and edge of each piece, and label the length so that you know what goes where. Familiarize yourself with all the joints—the type and the dimensions.

2. Take one stile and use the pencil, rule, and square to lay out the position of the rails and the mortises on one face. Use the square to run the lines over to the back edge, and then mark a small extra allowance for the wedges. The mortises need to be the width of the groove and centered within the thickness of the edge.

3. When you are happy with the markings, set the two stiles together and use the pencil and square to run the lines over on all the edges, both back and front.

4. Set the mortise gauge to a mortise chisel that is approximately one-third the width of the wood (say ⅜ inch). Then center the gauge on the thickness of the stile and gauge the mortise holes on the side face.

5. Select one rail and use the same setting on the mortise gauge to lay out the thickness of the tenons. Use the square to establish the position of the shoulders.

Cutting the Mortises

1. To cut the mortises, use the chisel and mallet in much the same way as for cutting regular mortises. First chop a small V-cut at the center and then back the chisel out from the center, all the while chopping deeper and deeper cuts at about ¼-inch intervals. Stop short of the squared line, and then turn the chisel around and repeat the procedure on the other side of the central V-cut. Clear the chips out with the chisel and then repeat the sequence for the next level of waste. When you have cut about halfway through the stile, turn it over and chop in from the other side. When you have chopped out the mortise from both sides of

48 • *Doors*

the stile, carefully pare the remaining waste back to the gauged line. Repeat this whole procedure for all the mortises—single and double. When cutting the mortises for the muntins—on the inside of the top rail, on both sides of the lock rail, and on the top side of the bottom rail—follow much the same procedure as already described, only this time around, the mortises need to be cut shallow and blind.

2. Use a ⅜-inch paring chisel to taper the ends of the mortises on the outside of the stiles by about ¹⁄₁₆ inch to allow a taper for the wedges.

Ploughing the Grooves

1. When you have completed all the mortises, use the plough plane to cut all the grooves on the inside edge. You might use a wooden plough plane, a dedicated metal plough plane, a combination plane, a router, or one or two other tools. It is important that the ploughed groove fall within the mortises and be the same size as the tenons. The groove must not be allowed to wander even a little to one side.

2. While the plough plane is in hand, run a groove through a small piece of off-cut waste. The resultant block is called a mullet, and it is used as a gauge to test the edge of the panels for a good fit. Some woodworkers use a piece cut from one of the horns for this task.

Cutting the Tenons

1. After you have cut the grooves on the appropriate sides of all stiles, rails, and muntins—all to a depth of about ½ inch—next comes the task of cutting the single tenons on the top rail and the paired tenons on the lock and the bottom rails. The method for cutting the single haunched and the paired single tenons is much the same as for cutting ordinary shouldered tenons, the only difference being that after you have cut the basic tenon, you need to modify it for the haunch or the paired form. After laying out tenons with the square and the gauge, the procedure is to score the shoulder line deeper with a paring chisel, set the rail flat down on the bench and hard up against the bench hook, and

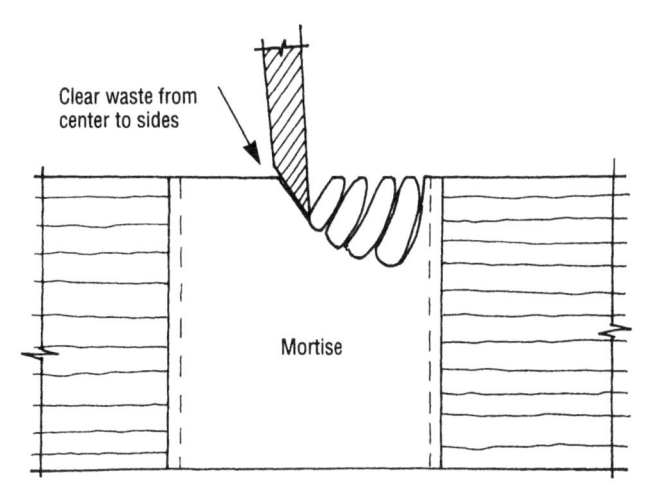

A cross section through the stile to show the mortising procedure. Once you have chopped out from the center, go back to the center and scoop out the chips of waste. With the chips out of the way, turn the chisel around and repeat the procedure for the other side. When you have cleared the waste and broken through, carefully pare back to the squared line, as shown by the broken line.

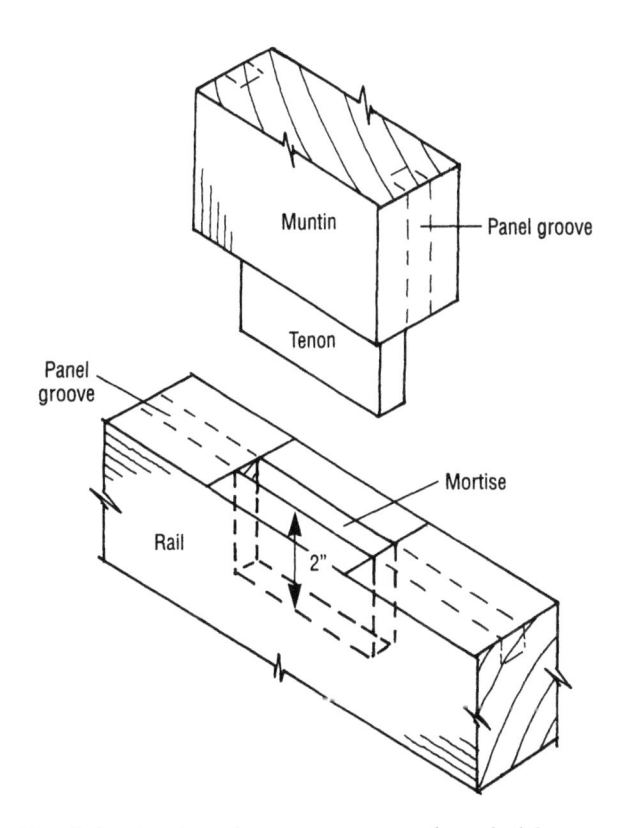

Detail showing the stub or stump tenon at the end of the muntin. Note the position of the panel grooves.

Making a Four-Panel Door • 49

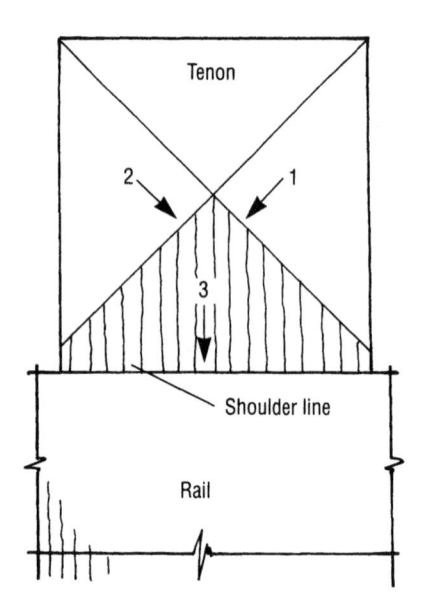

The order of kerfs for cutting a tenon: The first cut establishes one side and the top, the second cut establishes the other side, and the third and final cut runs down to the shoulder line.

> **SPECIAL TIP**
> There are many ways of cutting haunched tenons. Some woodworkers use a saw to cut the cheek waste down to the level of the top of the haunch and then set the rail flat down on the bench and use a rabbet plane to form the haunchings and the roots of the tenons. Other woodworkers believe that the whole operation is best done with a saw.

saw the shoulder down to the gauged tenon line. Do this on both sides of the rail. Set the rail in the vise so that the tenon is uppermost but tilting away, and use a half ripsaw to remove the cheek waste. Do this with three cuts. The first cut makes the kerf that guides the second cut, and then the two cuts act together as a guide for the third cut. Repeat the procedure for the other side. This procedure will have to be modified somewhat for the paired tenons—because of the width of the rail—but it is much the same. Some woodworkers prefer to cut the shape of the tenons and haunchings by clearing the between-tenon waste before they clear the cheeks. They say that this makes it easier to saw and gauge the thickness of the tenons.

2. Take the middle rail and—remembering to leave a haunch that is as long as the depth of the groove—use the backsaw and the coping saw to clear the waste from between the two tenons. Do this on both ends of the rail.

3. Repeat the procedure on the bottom rail, only this time around, leave a haunched shoulder on the bottom side.

Fitting

1. When you have cut all the tenons and trued up all the shoulders with a shoulder plane—the haunch shoulder on the top rail, the haunched paired tenons on the middle lock rail, the muntin tenons, the paired tenons with the haunching, and the shoulder on the bottom rail—then you need to trim everything to fit. The best way to proceed is to set the frame up loosely, see how the joints come together, and then generally test for squareness.

2. Cut the panels to size, so that there is an all-around clearance of about $1/16$ inch, and then use

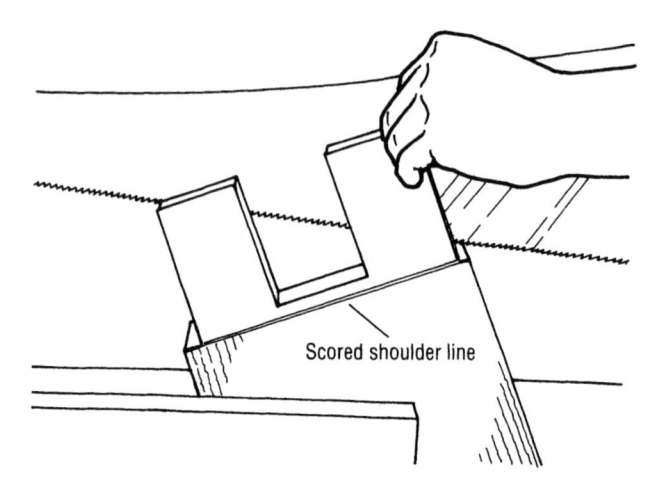

Sawing the cheeks on the paired tenons. The rail is secured at the correct angle in the vise, the saw is steadied with the thumb to the side (to keep the kerf to the waste side of the line), and the cuts are made in much the same way as for a single tenon. Note that in this instance, the between-tenon waste has been removed before the cheeks.

50 • Doors

the smoothing plane and the mullet gauge to bring them to a good fit and finish. Have a trial fitting of the panels within the frame.

3. Having achieved a good fit, knock the door loosely together and stand it in or near the room where it is to be hung for what is termed a second seasoning. The length of the second seasoning depends on the condition of the wood, how urgently you need the door, and your patience. Old-time woodworkers reckon that anywhere between three weeks and six months is fine, but two weeks should suffice.

4. Knock the door together for a final testing, trim to fit, and then disassemble for gluing.

Assembling and Gluing

1. Cut twenty small wedges, have a dry-run assembly so that you are well prepared and everything is comfortably at hand, and then start gluing. Smear glue on the tenons and shoulders of the muntins and rails, make sure that the panels and the ends of the rail tenons are completely free from glue, and then clamp up the door on a level surface. Make sure that the whole works is level and square.

2. Dip the wedges in glue and drive them home. First wedge the top of the top rail to drive it down and the bottom of the bottom rail to drive it up, so that the muntin shoulders are hard up against the middle rail. Then follow with the other wedges.

Fixing the Moldings

1. This particular door has a generous bolection molding on one side and a plainer panel molding on the other. You need to decide whether to make your own moldings or buy them. Of course, much depends on your tool kit—whether you intend to use old wooden molding planes, a combination plane, a router, or whatever.

2. When you have obtained the molding—you need 54 feet for each side of the door—use the miter block and the backsaw to cut it to size. The bolection molding needs to be cut so that it laps over the edge of the frame.

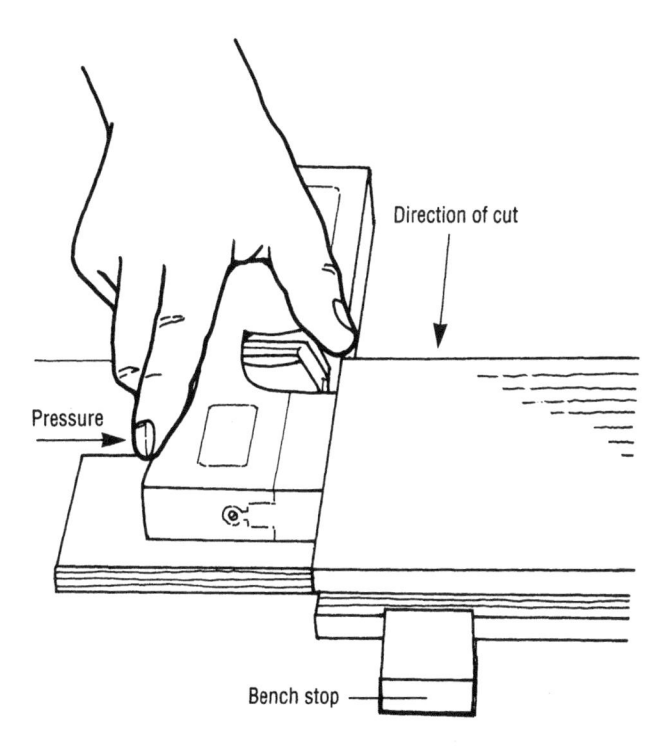

Shooting a shoulder is the procedure of setting a shoulder plane flat down in its side—so that it is hard down on the tenon and hard up against the shoulder—and then running it through to true up the surface.

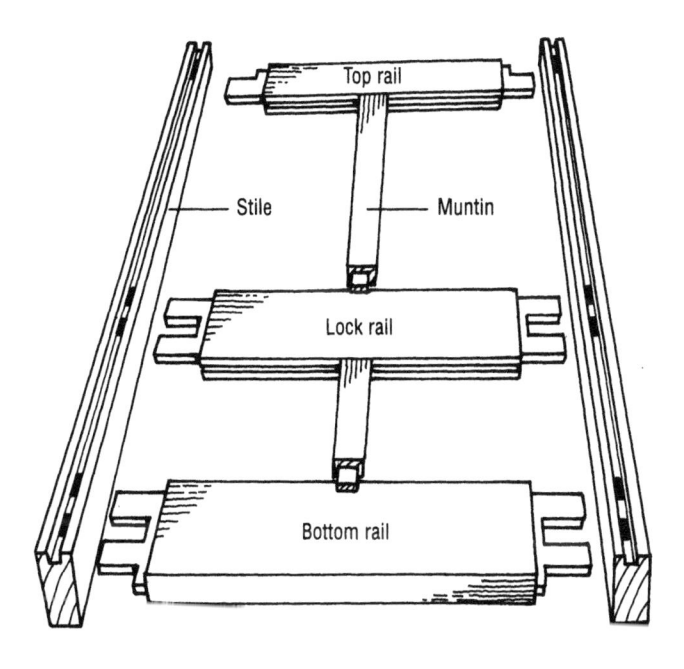

The frame is set out on a level surface just before testing the fit. Note that the near ends of the muntins are cocked up so that you can see what's going on.

Making a Four-Panel Door • 51

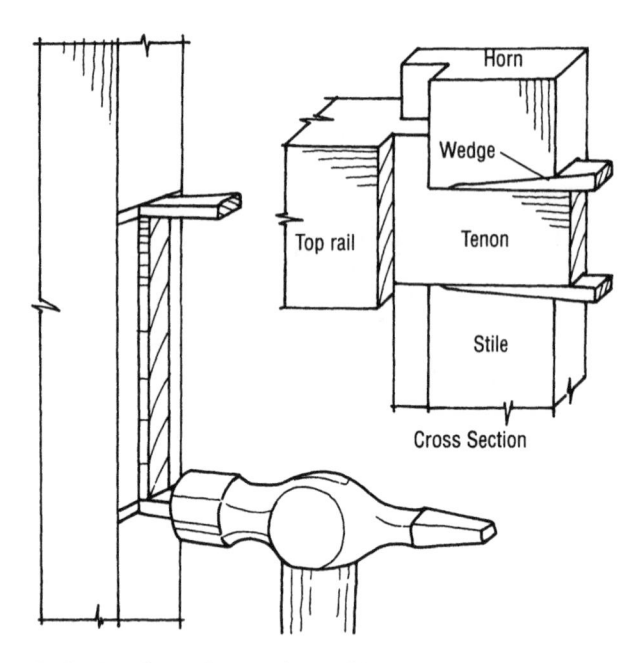

Left: Dip the wedges in glue and hammer them home.
Right: cross section showing the wedges in place.

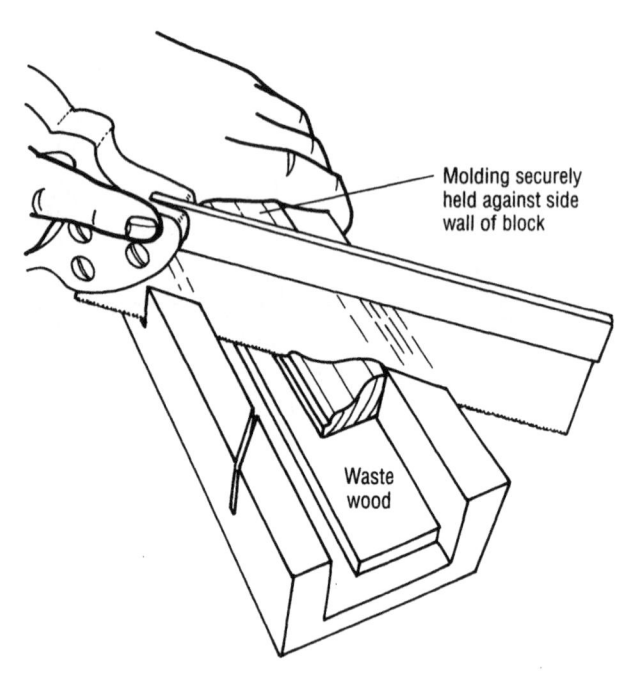

Using the miter block to cut the miters on the moldings. The piece of waste wood under the workpiece prevents damage to the block.

3. Having cut and trimmed the molding to size, next comes the tricky task of nailing it in place. It's tricky because the molding must be nailed only to the frame—not the panels—and it must also be fitted in place without the hammer bruising the wood. This is best achieved by placing a thin sheet of metal between the hammer and the panel and then sliding the hammer to its mark. Some woodworkers use the blade of an old square for this task.

SPECIAL TIP
To clarify the statement that the molding must be fixed only to the frame, it may have been more correct to say that the molding should be fixed only to the frame or only to the panel—never to both. Some woodworkers opt to screw the bolection molding in place on the panel—with screws coming through from the other side—and then cover up the screw heads with the plain panel molding that is nailed on the frame at the other side of the panel. Either way, the important thing is that the panel be free to move within the frame.

4. With the molding in place, the whole works needs to be put in order. Punch the nail heads below the surface and fill the holes. Cut off the ends of the tenon wedges and the horns at the top and bottom of the frame. Then rub the whole works down to a smooth finish, and paint or otherwise finish the door to suit.

AFTERTHOUGHTS AND OPTIONS
- The order of the steps presented here is only a suggestion. There are, of course, many ways to proceed. Some door makers work the mortises and tenons first, then plough the grooves and rabbets, and finally cut the joint shoulders to fit. The best option if you are a beginner is to first follow the steps as described, and then find your own way when you're more experienced.
- Although doors can certainly be made more swiftly with routers, spindle molders, and the like, hand tools are best if you intend to make only a couple of doors.

- If you want to make your doors completely by hand—all the joints and everything else—you need to search for some good old tools. Flea markets are good places to start. You need a wooden plough plane with a set of nine irons or cutters. This tool cuts grooves from ⅛ to ⅝ inch wide and to any depth down to about 1¾ inches. Such a plane runs the grooves straight over the mortises without stopping or jolting and is fitted with a fence and a depth stop. You also need a set of wooden molding planes and a couple of well-chosen heavy-duty mortise chisels.

- If you want to really enjoy yourself, you need a metal combination plane of the Stanley 45 type. Such a plane can be used for cutting everything from rabbets and grooves to tongues and grooves and simple molding.

- When choosing your wood, spend time making sure that you get the best. It's no good if the stiles have knots near the mortises or grooves, or if there are end splits that run up the length of the stiles. For the panels, which are generally 10 to 12 inches wide for four- and six-panel doors, look for wood that comes in wide, stable widths that can be cut into thin sections. You might try yellow pine, redwood, Parana pine, American oak, or beech. If you want to make a low-cost interior door that will be painted, you can use best-quality multicore plywood for the panels.

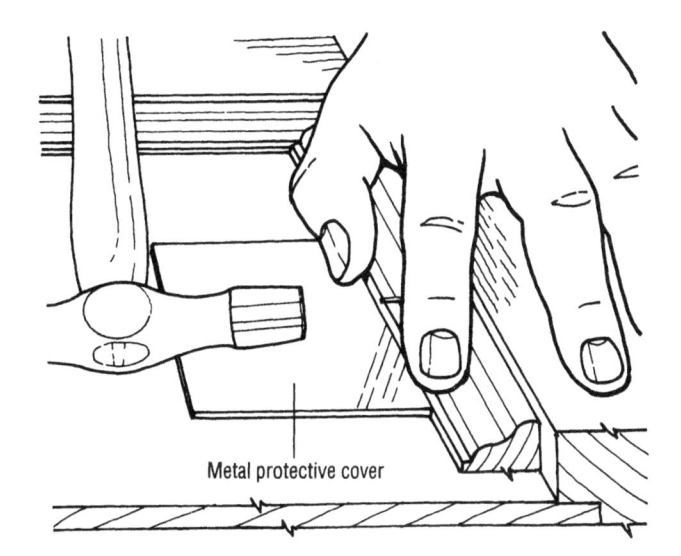

Securing the moldings. Set the moldings in place against the frame, protect the panel and the edge of the molding with a sheet of metal or thin plywood, and tap the nails home. Some woodworkers use a flat-sided glazing hammer for this procedure.

- Because the horns at the ends of the stiles protect the door from damage when it is being moved around the workshop, they need to be left on until the last moment.

- Be sparing with the glue. Don't glue the panels into the frame, or the moldings to the frame or the panels, or the ends of the rails to the stiles.

Making a Four-Panel Door • 53

Hanging and Fitting a Door

If you go into an English pub on a cold winter evening, when the icy wind is blowing and the rain is slicing down like iron stair rods, a shout will immediately go up: "Put the wood in the hole"—meaning, "Would you be so kind as to close the door?" A door is a door is a door—but only if it is nicely hung on hinges so that it fills the door frame, and only if it is fitted out with suitable hardware that allows it to be easily opened and closed.

Depending on the shape, size, and character of the door, there are all manner of hinges, catches, and door framing options. When you are ready to hang the door, you first need to consider the hole into which it is going to be hung—the actual door frame. Then you have to consider the size, shape, and fitting of the hinges, how you want the latch or catch to be fitted, and so on. It's easy when you know how. The following descriptions and step-by-step project will show you the way.

THE DOOR FRAME

Exterior door openings and vestibule openings are usually framed with a good-sized solid section, with all the rabbets and molding being worked from the solid. The vertical members of the frame are called *posts* or *jambs*; from top to bottom, the horizontal members are known as the *head* and the *sill*. If the door frame is extended vertically so that there is a fan light between the top of the door and the head of the door frame, the horizontal bar just above the top of the door is known as the *transom*.

Exterior doors are usually hung so that they open into the house. The good-sense reason for this convention is that if the door were to open outward rather than inward, every time you opened the front door to receive guests, you would knock them off the doorstep. The only exceptions to this rule are stable-type doors that are fitted so that they open outward. The doorposts are tenoned into the head of the frame, so that the resultant horns on the ends of the head can be built into the wall.

Interior door frames—variously termed frames, casings, or linings—are usually built up from a number of relatively slender parts. The character of the lining is determined, to a great extent, by the thickness and structure of the wall. For example, if the wall is made from wooden stud work, the opening might well be lined with ¾- to 1-inch-thick wood, with a plain casing nailed on to cover the joint between the plaster and the jamb; the doorstop could be made by nailing on a bit of 1½-by-½-inch pine or another inexpensive wood. For a more substantial brick wall, the lining is usually thicker—about 1½ inches thick—with the doorstop being rabbeted from the solid. In this instance, the lining can be worked with either a single or a double rabbet—the idea being that the door can be hung on either side of the wall thickness. If the brick wall is 14 inches thick, the opening might well be framed and paneled with stiles and rails, with the framing rabbeted so that the door can be fitted on either side of the opening.

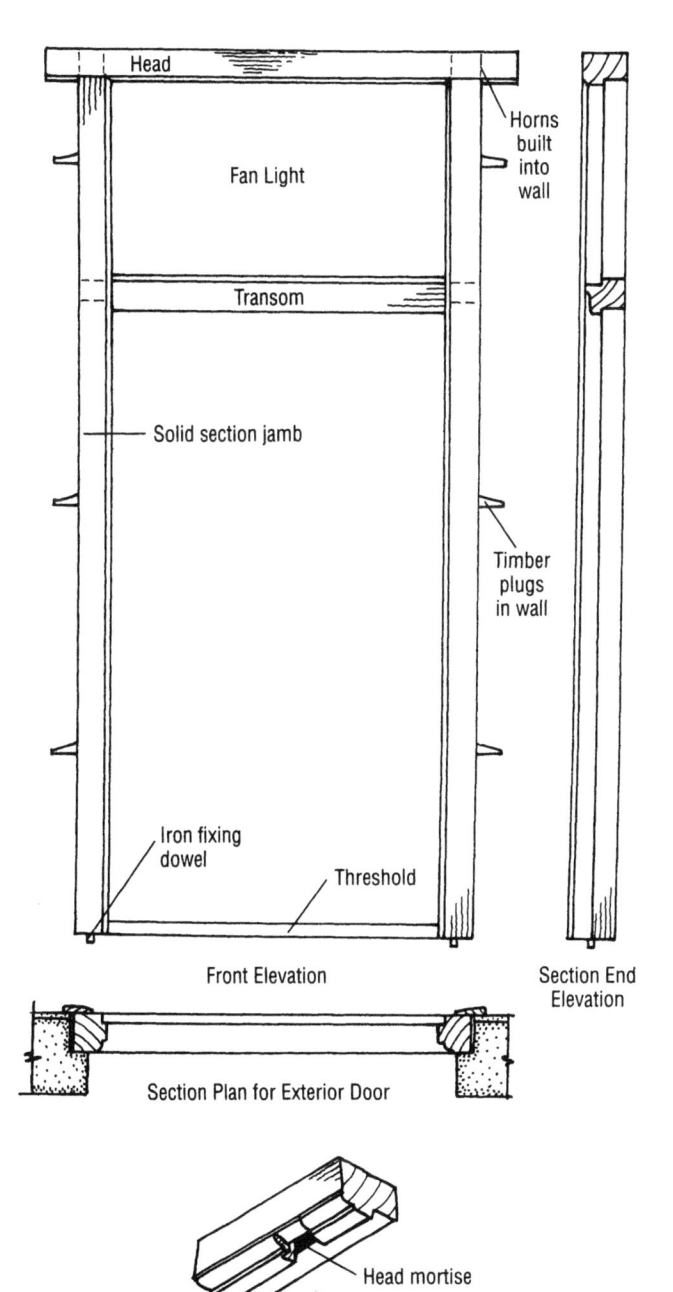

The architraves are molded sections nailed to the room side of the lining. They function on two levels: They cover the joint between the woodwork and the plasterwork, so that you can't see the movement that occurs between the wood and the plaster, and they provide ornamentation. As a general rule, the thicker the wall and the heavier the door, the wider and more fancy the architraves.

HINGES

Good door hinges are like the Victorians' notion of good servants—they are all about us, they never cease to work, yet they are rarely seen and never heard. There is no denying that hinges are vital to the smooth functioning of a door. Most of us become aware of the door hinge only when it starts to squeak. But if you want your doors to open and close without fuss—no squeaking, shuddering, jamming, dropping, or otherwise complaining—you must buy the very best hinges and fit them properly. Although the choice of hinges is huge, there are only two basic types: those that are recessed into

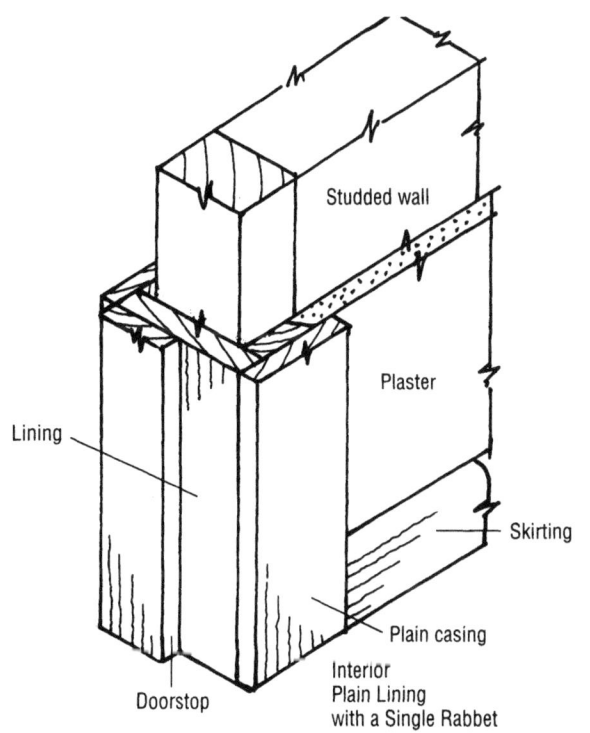

A classic solid-section door frame for an exterior door, showing a front elevation, a section plan, and a section end or side elevation.

Top to bottom: *Horns are built into the wall, timber plugs are driven into the joints of the wall—meaning between the bricks or stones—and iron dowels are driven into the stone sill.*

A section through a plain door lining, used for basic dry-lined walls and for block and plaster. Note how the plaster is covered and contained by the casing.

Hanging and Fitting a Door • 55

Interior Double Rabbeted Plain Lining

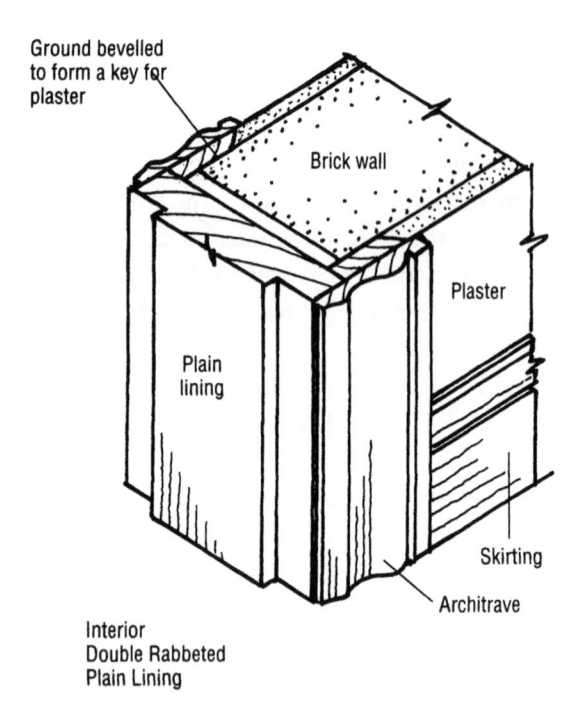

A solid wood lining for a substantial brick wall. Note the architrave and the skirting with the decorative molding. The good thing about this option is that the door can be hung on either rabbet—it gives you a choice.

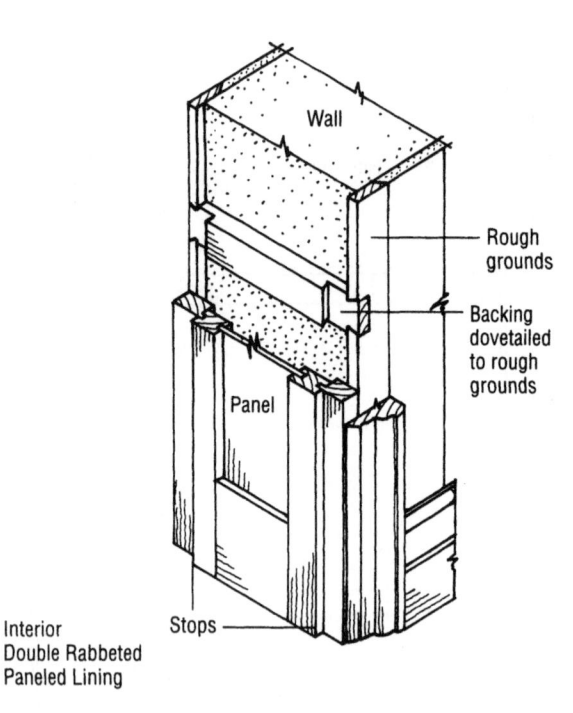

Interior Double Rabbeted Paneled Lining

A top-quality lining for superior work—meaning a good double-thickness wall. The lining is fixed to wooden battens or grounds that are held in place by dovetailed backing strips. This option allows you to fix the door on either rabbet. Sometimes—such as for adjoining bedrooms—two doors are fitted.

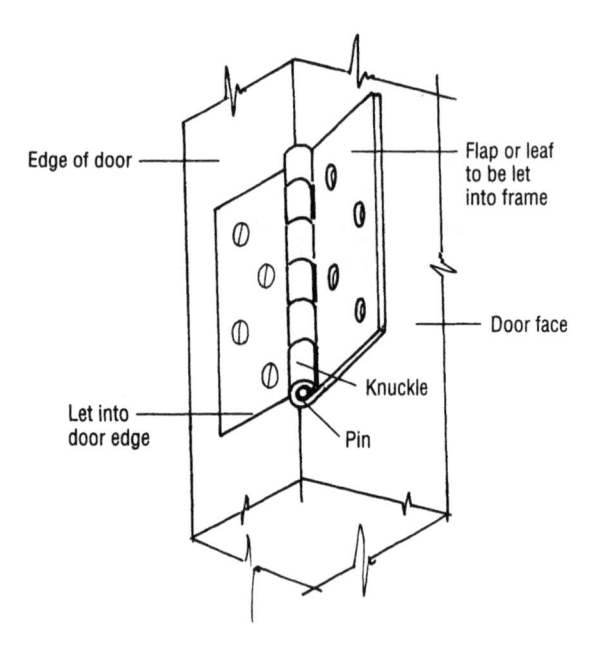

A straight butt hinge driven into the edge of the door. The rolled knuckle indicates that it is pressed rather than cast.

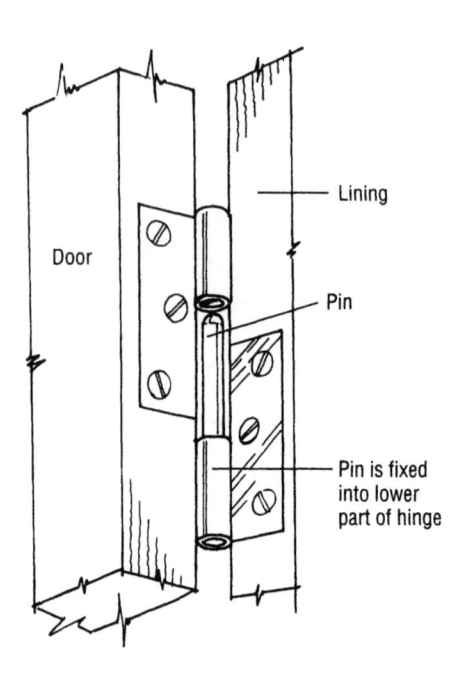

A lift-off butt hinge. The half of the hinge that contains the fixed pin needs to be fixed to the lining, with the pin pointing up.

a mortise, and those that are attached to the surface.

Straight Butt Hinges
Straight butt hinges are the type used on most panel doors. They might be made from anything from cast brass or bronze to pressed steel or cast iron. Brass hinges are usually fitted to exterior doors, pressed steel hinges are fitted to lightweight interior doors, and cast-iron hinges are fitted to heavyweight interior doors. The flaps or leaves are mortised—one into the jamb and the other into the side of the hanging stile.

Lift-Off Butt Hinges
Lift-off or heave-off butt hinges are, as the name suggests, designed so that you can easily remove the door without unscrewing the hinges. For example, if you need to remove the doors and shutters for the summer months, lift-off hinges would definitely be the answer. There's no messing around with screws; you just lift off the door, and the task is done. These hinges are set in mortises, as with regular butt hinges.

Knob-Pin Hinges
Sometimes known simply as pin hinges, these are designed in much the same way as most regular butt hinges. The only difference is that the pin extends out from the knuckles, where it is topped off at each end with a decorative knob. Most knob-pin hinges are made from brass. However, some are made from blued steel, with the visible part of the hinge being fretted and chased with decorative motifs. Hinges of this type look good on heavy, ornate interior doors—the sort of fancy doors you see in grand houses.

Rising Butt Hinges
Rising butt hinges are designed so that the wearing surfaces of the leaves are helical; thus when the door is opened, it rises on the hinge. The idea is that when the hinge rises, the door clears the carpet. When the door is released, it slowly falls back into the closed position.

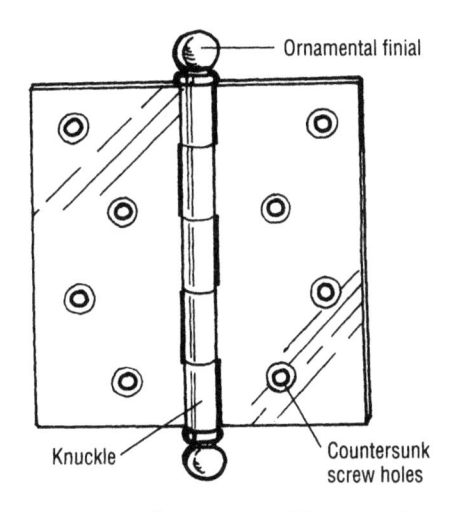

A knob-pin hinge. Note the ornamental finials on the ends of the pin. Sometimes the finials are threaded so that they can be unscrewed for easy removal and maintenance.

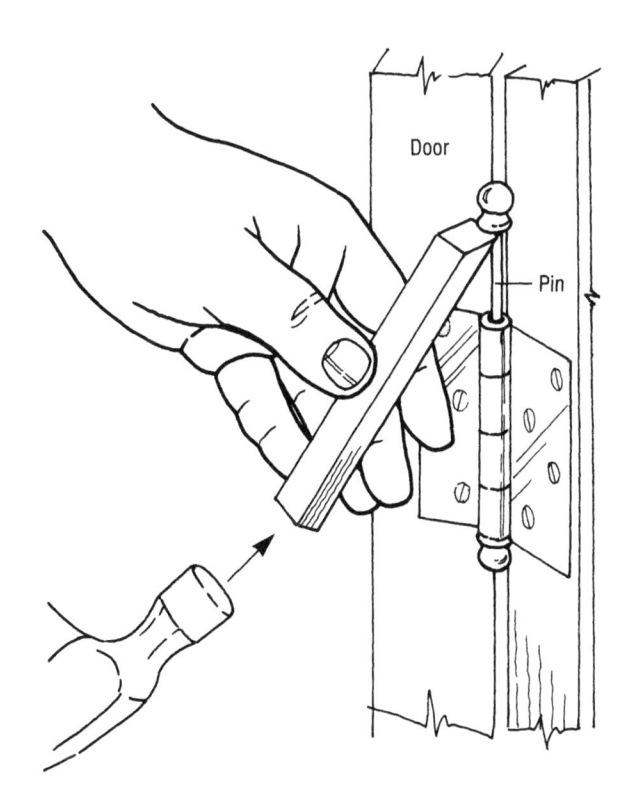

In this instance, the bottom finial is fixed to the bottom knuckle, and the top finial is fixed to the pin. To remove the pin, you turn the top finial and unscrew the pin from the bottom finial; then you use a strip of hardwood and a hammer to tap the pin clear. The best procedure for removing the door is to get a helper to hold and support the door while you remove the pin from the bottom hinge.

Hanging and Fitting a Door • 57

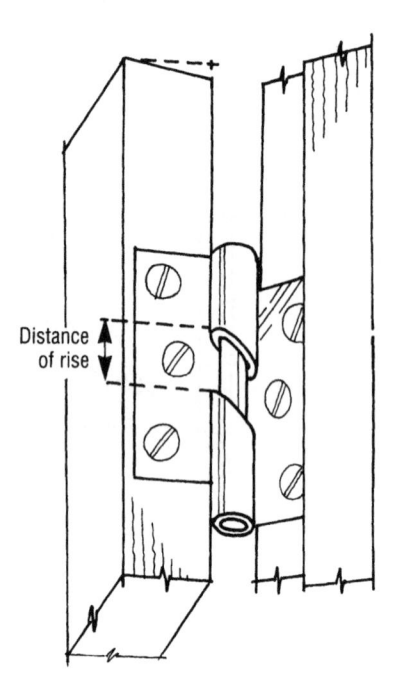

Rising butt hinge. Although more clearance is needed at the top of the door because of the rise, the problem of the door fouling the frame can be avoided by slightly beveling the top edge of the door.

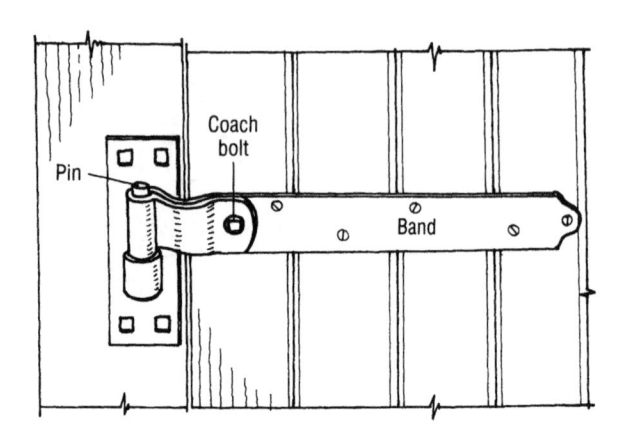

A heavy-duty strap hinge designed for use on large, heavyweight doors—ideal for stables and workshops. If you are worried about security, you can reverse the top hook or pin plate so that the pin is pointing downward and the door can't be lifted off. Note that the coach bolt runs through the foldover of the strap and on through the door stile.

Strap Hinges

These heavyweight hinges are used on large external doors, where the primary need is for strength. They are bolted in two parts—onto the face of the door and onto the jamb—so that they can be lifted off. Such hinges are perfect for heavy workshop and stable doors.

Cross-Garnet Hinges

Sometimes known simply as tee hinges, these hinges are used mostly on cottage and country doors of the batten and ledged type. There are many different types, from the ordinary flat, pressed steel hinges that can be found in most hardware stores to the beautiful items seen on antique doors. Such hinges are still being made in the old traditional patterns by many art blacksmiths, and it is still possible to get authentic hinges. There are

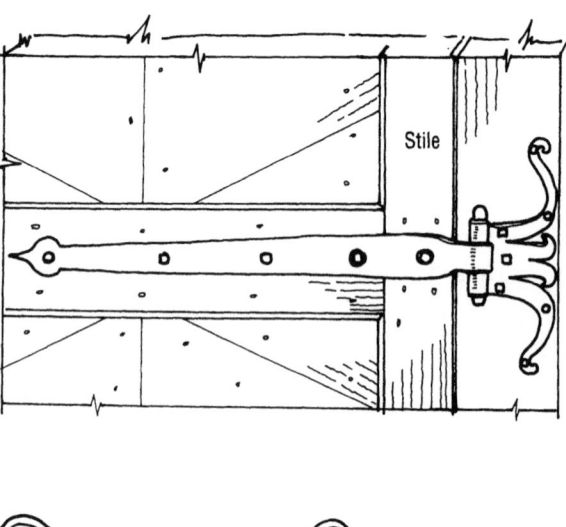

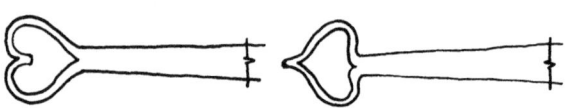

Cross-garnet hinges come in all shapes and sizes. The ones shown are prime examples of the blacksmith's art. Not so long ago, such hinges were thought to be a bit primitive, but they are now considered folk art.
Top: *A hand-forged wrought iron hinge with a ram's horn detail on the pin plate and a heart motif on the end of the strap.*
Bottom: *Two traditional split-end designs that draw their inspiration directly from the making techniques. The iron is hammered, split, rolled, and welded.*

many county and regional styles; some have the end of the hinge strap beaten out to make a heart form, or the end of the strap divided into two forks and then rolled back to make an open heart. Although these hinges can be fitted on either side of a batten door, they must be aligned so that they are screwed or bolted through to the ledges.

H or Parliament Hinges

These hinges are H-shaped, with the knuckle occurring on the crossbar of the H. Such hinges are usually used on period-type cottage doors, with the flaps mounted on the surface of the door and the jamb. H hinges are made of forged iron for good, solid exterior doors and of brass for fancy interior doors.

L hinges

Also called quadrant hinges, L hinges are much the same as H hinges, the only difference being the shape of the flaps. The L-shaped part of the hinges is screwed to the door. They come in left and right configurations.

Ram's Horn Hinges

Ram's horn hinges are sometimes described as Pennsylvania German or even American folk-art hinges. These hand-forged iron hinges are still

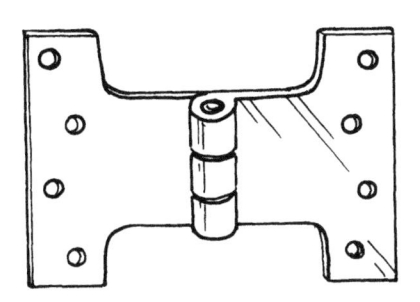

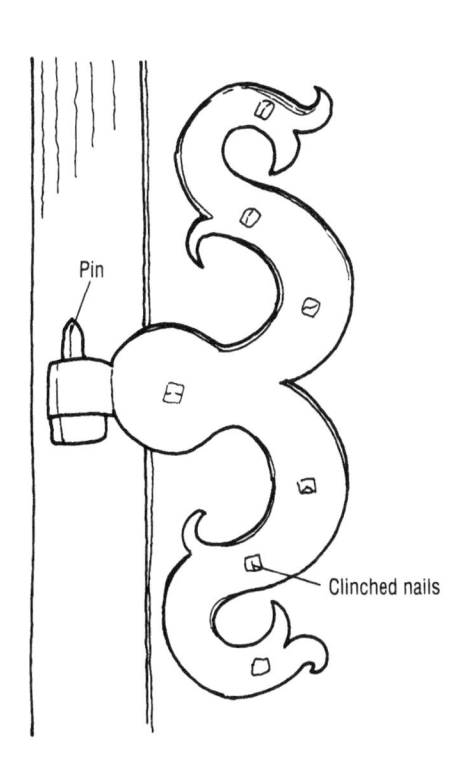

A parliament or H hinge. Although such hinges are sometimes fitted on the surface of the door and the jamb—so that both flaps are in full view—they can also be fitted in much the same way as a regular butt hinge. The idea is that the protruding knuckle allows the door to swing clear of surrounding architraves and skirting.

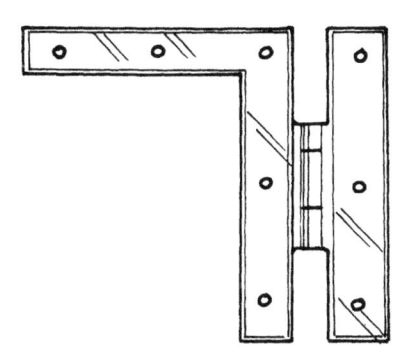

A classic L hinge. These types of hinges are found on Early American doors and shutters from the seventeenth to the late nineteenth centuries. The flaps are mounted directly on the face of the door and the surround, the idea being that the shape helps to strengthen and support the frame of the door where the rail meets the stile.

A traditional ram's horn hinge made from forged iron. The pin-hook is spiked into the door frame, and the ram's head is dropped onto the pin. The plate is punched with square holes to take square-section rose-head nails. Although this particular hinge is from a Pennsylvania German house, such designs were common all over. In some coastal districts in England and along the U.S. East Coast, the ram's horn detail is modified so that it looks more like a fish's or mermaid's tail.

Hanging and Fitting a Door • 59

made by country blacksmiths in the United States, England, Germany, and elsewhere. The design and form of the hinge are more or less the direct, almost inevitable result of the forging process. The hot iron is looped, twisted, scrolled, and hammered with a minimum of fuss, and the ram's horn form is the result. Although the design illustrated is symmetrical, other hinges of a similar type are worked so that they are left- or right-handed to fit the corners of the door.

LATCHES, LOCKS, HANDLES, AND OTHER FITTINGS

All doors need a handle so that you can pull them open or closed and some sort of lock, latch, or catch to hold them closed. There are hundreds of different forms and models to choose from—in association with other fittings such as doorplates, mail slots, and the like—but there are just a handful of generic types. There are knobs, thumb latches, rim locks and latches, bars, bolts, mortise locks, and various ultra-modern Yale-type deadbolts. The following listing gives you some idea of the possibilities.

Thumb Latch

Also known as a Norfolk latch, Suffolk latch, and a number of other names, this fastener is designed specifically for use on thin batten and ledged doors. Most latches of this character consist of a plate or carrier with an integral bridge-shaped handle and a hinged lever that runs through the thickness of the door to engage with the latch. The handle plate is always mounted on the outside of the door—that is, the swing side on which the door opens. The handle part is gripped while the thumb of the same hand presses down on the lever. This action causes the latch to disengage from the catch. The good thing about a unit of this type is its simple, last-forever structure and the fact that it is very easy to fit.

Tower Bolt

The tower bolt is a horizontally mounted bolt or bar that slides or shoots to become engaged in a staple fitted into the frame. Fitted with screws and/or bolts to the inside of the door, a fastening of this type is inexpensive and makes for a good, secure door, especially if you have bolts on both the top and bottom of the door.

Rim Lock

The rim lock serves as both a latch and a lock. The latch is turned and operated by means of a square spindle operated by a pair of knobs or handles. A key is used to operate the lock or bolt. A box is screwed to the frame of the door to receive the bolt. The good thing about this type of lock is that the whole works can be screwed onto the surface of the door and the frame. All you have to do is bore a hole through the door for the square spindle. Rim locks are made from pressed steel and/or brass; some fancy types have a box that is both fretted and

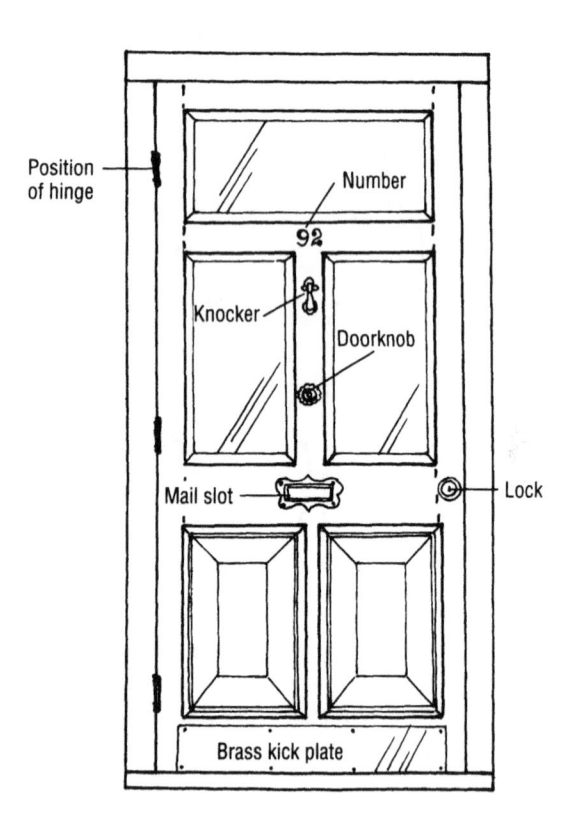

The traditional arrangement of the door fittings on a classic five-panel entrance door.

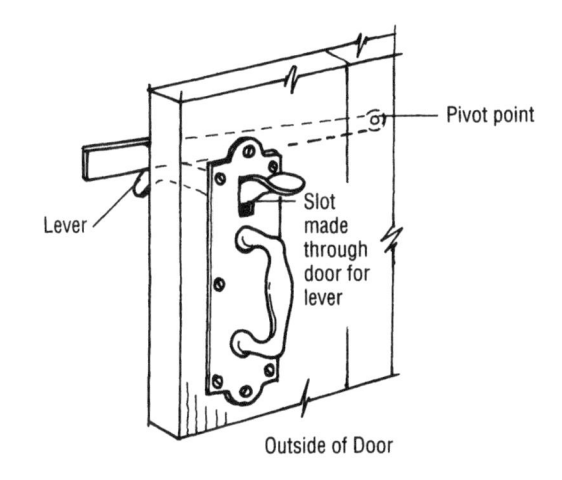

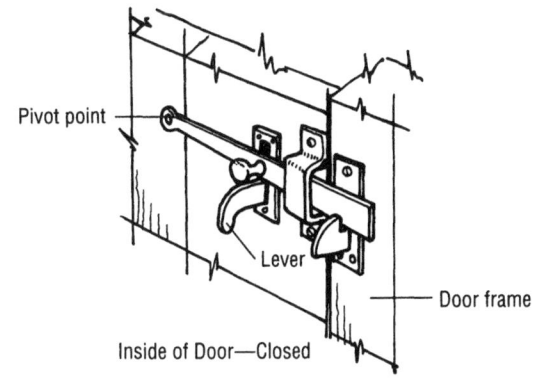

Known variously as a thumb latch, a Norfolk latch, and many other names, this latch is a good choice for country doors. If you are looking for something special in the way of design, visit an architectural salvage company and search out latches made at the beginning of the nineteenth century.

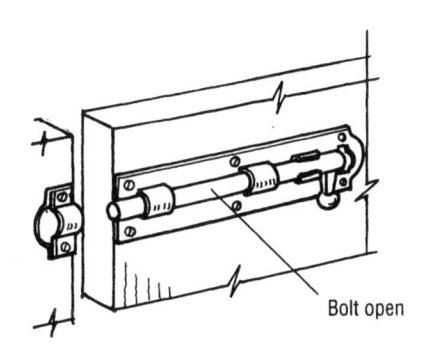

The tower bolt is a winner on many counts: It is strong, easy to fit, and inexpensive. Try to avoid the fussy designer options with fancy doodads. It's much better to stay with the simple, traditional cast-iron and pressed steel form. If you want extra security, use heavy-duty coach screws (screws with square or hex heads).

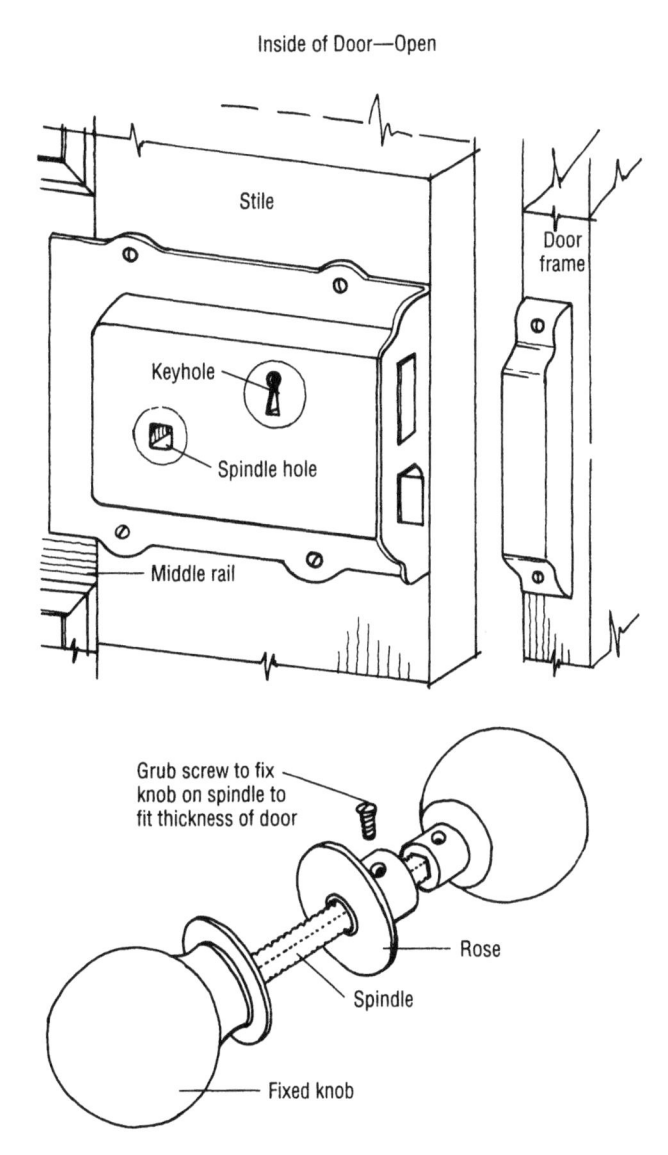

If you want a traditional, inexpensive, easy-to-fit lock on a low-security door, the rim lock is a good first choice. If you want something fancy, it is possible to buy salvaged locks from old Victorian and Edwardian doors. They look much the same as the pressed steel variety, but everything except the springs is made of brass.

With this particular design, the spindle, complete with one doorknob and rose, is slid into place through the lock. The second rose is set on the spindle and brought up to the other face of the door. The other knob is threaded on the spindle and adjusted to the thickness of the door, then the second rose is aligned so that its screw hole is directly over the screw hole of the second knob, and the grub screw is tightened to clinch the knob on the spindle.

Hanging and Fitting a Door • 61

chased. If the rim lock has no latch, it is termed a "dead lock," meaning that it can be opened and closed only by means of a key.

Stock Lock

The traditional stock lock is a bigger and fancier version of the rim dead lock. It is usually in the form of a large wood and metal box that is mounted to the face of the door, with the box being decorated with brass or steel corner plates.

Mortise Lock

Although the mortise lock serves the same purpose as the rim lock—it carries both a latch to hold the door closed and a lock to make the door secure—the main design difference is that the works are hidden in a mortise within the thickness of the door. Most modern panel doors are fitted with mortise locks. The main advantage of a mortise lock is that the face of the door is unencumbered; however, the fact that the lock is hidden away has resulted in the manufacture of poor-quality products that barely function. Whereas older mortise locks required that a square side mortise be cut into the side of the stile—not an easy task—many modern locks have cylindrical bodies or barrels that can be fitted by boring a hole.

Knobs and Handles

Doors are fitted with all manner of knobs and handles: bridge-shaped handles made from forged iron, turned wooden knobs, iron rings mounted on rose plates, integral pull-and-knocker handles, ball-shaped brass knobs, egg-shaped glass knobs, ceramic knobs, and many other designs. If you are looking for a special knob, visit an architectural salvage company and search out an old item. You can't beat some of the handles and knobs that were made a hundred or so years ago. If you do intend to use old fittings, make sure that they are complete and sound. Pay particular attention to the spindle, the rose that fits on the spindle between the door and

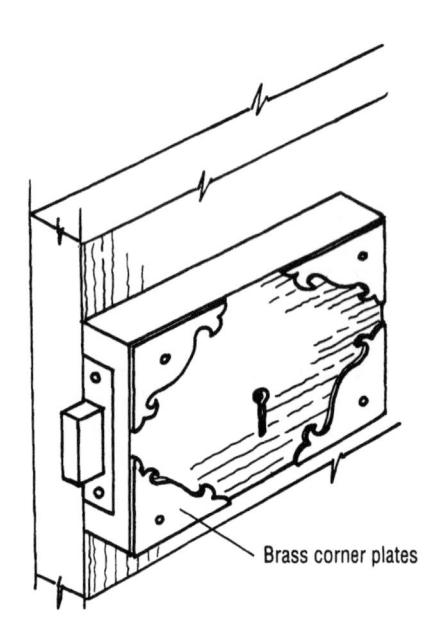

The stock lock was fitted to most early-nineteenth-century entrance doors. Although it certainly looks impressive—like everybody's notion of a prison lock—the workings are so crude that the lock can easily be picked with a couple of bicycle spokes. Nevertheless, some of the old locks are attractive—lots of wood and brass.

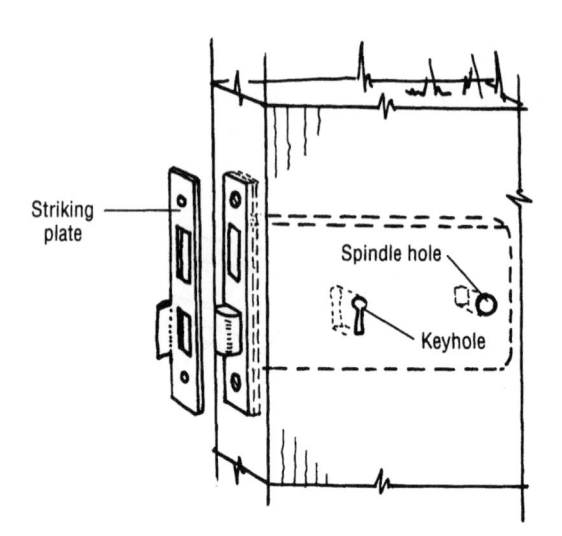

The mortise lock, in one form or another, is still the standard for most doors. Its only good feature is that it is hidden away in the body of the door. It does the job of holding the door closed, but for the most part, the action of cutting the mortise to fit the door weakens the fabric of the door.

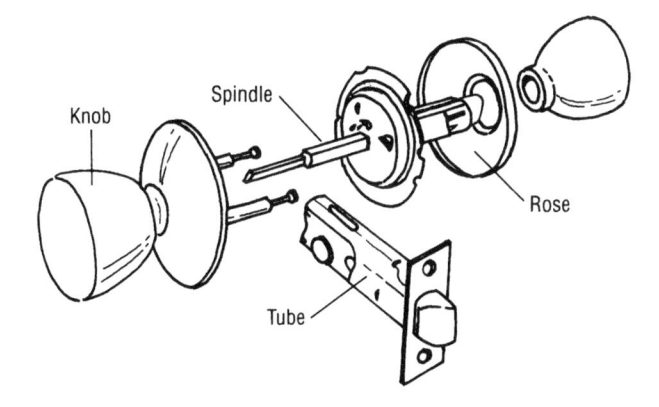

The modern cylinder lock is a good option when you want to fit a mortise-type lock. A plus is that the hole can be bored out rather than cut with a chisel. All you do is bore two holes—one through the thickness of the door for the spindle, and the other into the edge of the door for the cylindrical body of the lock. This particular model has a lock built into the body of the spindle, designed to disengage the spindle from the latch.

the knob or handle, and the little grub (or worm) screw that fixes the handle to the spindle (see the illustration of the rim lock).

Door Knockers

Traditionally, door knockers were mounted on the front door for the functional reason that they were used to inform the inhabitants of the arrival of visitors. They were also used as decorative status symbols. The bigger the house, the larger and more prestigious the front door, and the fancier the door knocker. To a great extent, the knocker has been replaced by the electric doorbell, but it is still used as a decorative statement-making feature. Knockers come in all shapes and sizes—cast iron lions' heads with rings in the mouths, brass Adam-style urns, forged Gothic knockers, brass figures mounted as knockers, brass wreaths, and so on. It is common practice to personalize the front door with a knocker that is in some way meaningful to the area. For example, in a fishing village, there might be knockers in the form of galleons, anchors, brass rings, ships bells, Turk's head knots, dolphins, and so on. And of course, there are knockers that relate to the family name or the name of the house.

For example, a family called Hart might have a stag as a knocker.

Mail Slots and Other Things

Some doors need a mail slot, and others need a number and a knocker. There is a tradition of trying to make the front door grand and special. Some people go in for lots of brass, others like carved nameplates, others like lots of color, and so on. The wonderful thing is that just about anything goes. If you have a fancy for carved trim, lots of brass, and gold linework, that's fine.

HANGING AND FURNISHING A PANEL DOOR

There is no single correct way to hang a door. Some woodworkers fit the lock and then the hinges; others fit the hinges and then the lock; still others always fit three hinges no matter the size and weight of the door. The following section describes one way to go about hanging a door, but there are

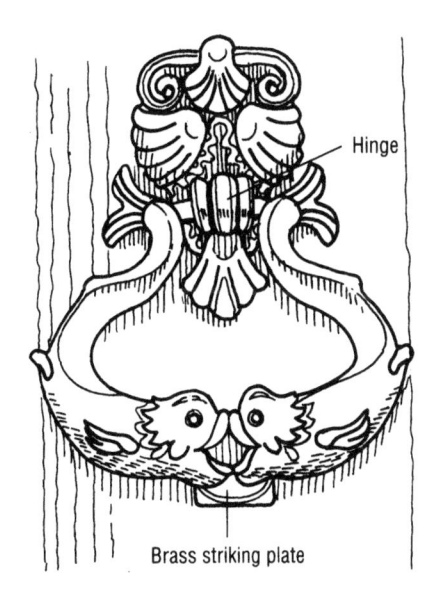

A traditional cast brass knocker—just the sort of thing you might find in a coastal town. This particular knocker extends the fishy theme by having the dolphins pivoted on a cluster of shells. So if your family name is Fish or Fisher, this is the knocker for your door.

Hanging and Fitting a Door • 63

many other methods. If you are a novice, follow the steps described here, and then go your own way when you've gained some experience. Although it is certainly easier to have a helper when you hang a door, the method described tells you how to do it on your own. The example used is an average-size panel door with a mortise lock and fancy ceramic knobs.

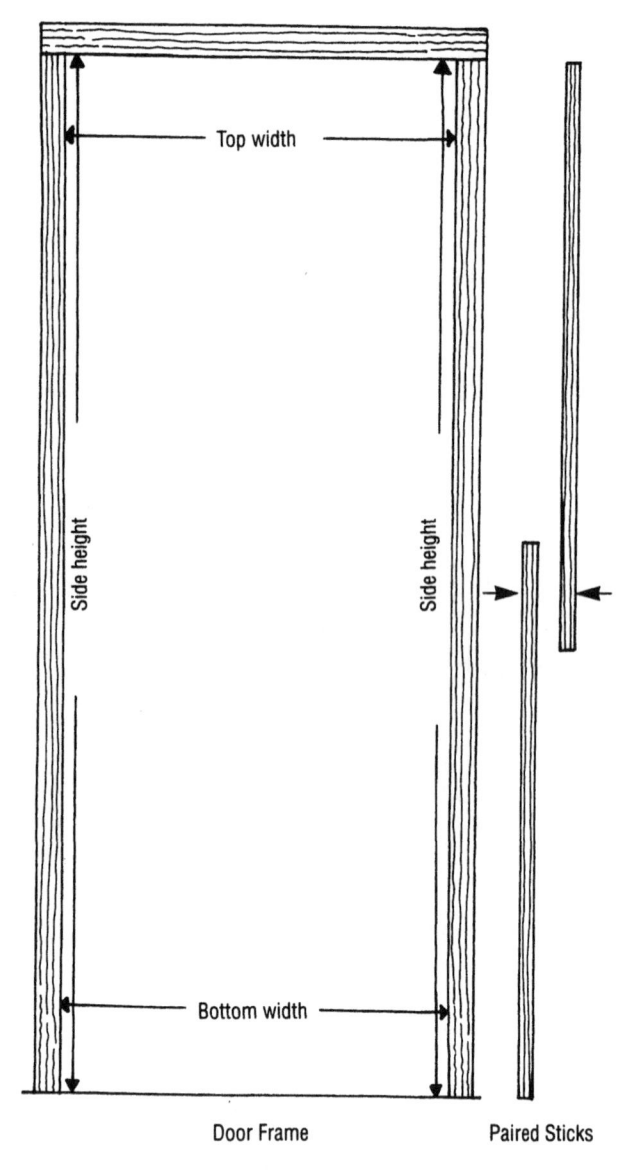

One of the easiest ways to measure a door frame is to use the traditional two-stick or paired-stick method. Clasp the sticks together—at the point shown by the two arrows—and then transfer this measurement directly to the door.

Tools
- A couple of sawhorses or wooden boxes, both at the same comfortable working height.
- A couple of thin battens (paired sticks) to transfer the height and width of the door frame.
- A crosscut saw.
- A square.
- A marking knife and an awl.
- A marking gauge.
- Two planes—a smoothing and a block.
- Two bevel-edged paring chisels—one that is as wide as or wider than the hinge flap, and one ¾ inch wide.
- A mallet.
- A hand drill with a pilot drill to fit your screws.
- A brace with a ½-inch-diameter long-shanked Forstner bit to fit.
- A screwdriver to fit the screws.

Materials and Fixtures
- A door that is fractionally wider and higher than the rabbet opening—that is, a new door complete with horns.
- A pair of butt hinges to suit the door. If the door is heavy, go for 4-inch hinges; otherwise, go for 3-inch. Use cast brass for exterior doors and cast steel for interior doors. Avoid pressed steel hinges.
- Countersink screws to fit the hinges, so that the heads are flush with the surface.
- A mortise lock to fit the design of the door.
- A pair of ceramic doorknobs with two rose plates and a spindle length to fit the thickness of the door—all with screws to fit.

Cutting the Door to Fit

1. With the door set across the two sawhorses or boxes, so that it's like a low table, use the paired sticks to take accurate height and width readings from the door frame rabbet, and then transfer the measurements directly to the door. Being mindful that the frame will almost certainly be less than true, take four measurements—both sides, the top, and the bottom.

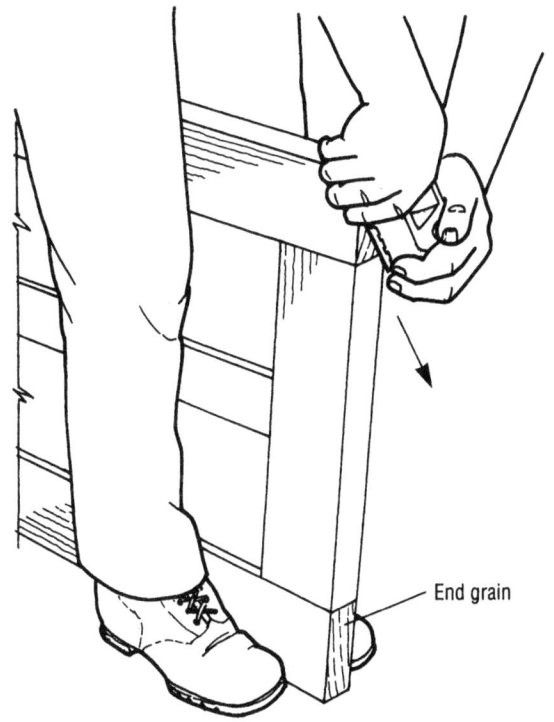

Saw off the horns and then plane the end grain to a smooth finish.

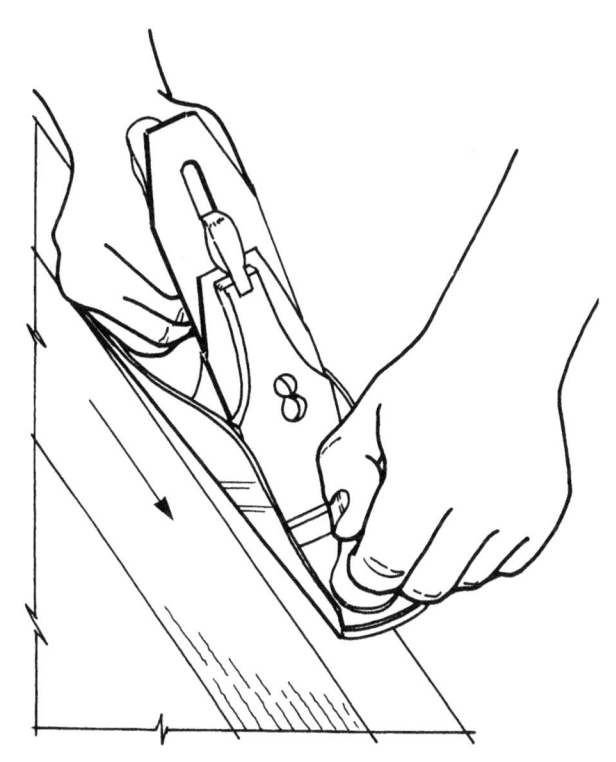

Use a smoothing plane to trim back the hinge or hanging stile.

2. After using a straightedge to connect the transferred marks, trim the door to fit. If the door is a shade too long, you need to cut back the waste. The best procedure is to saw off the horns first; then use the block plane to bring the ends of the door to a smooth, square finish. Run the plane from the sides to the center to avoid splitting off the grain at the ends of the stiles.

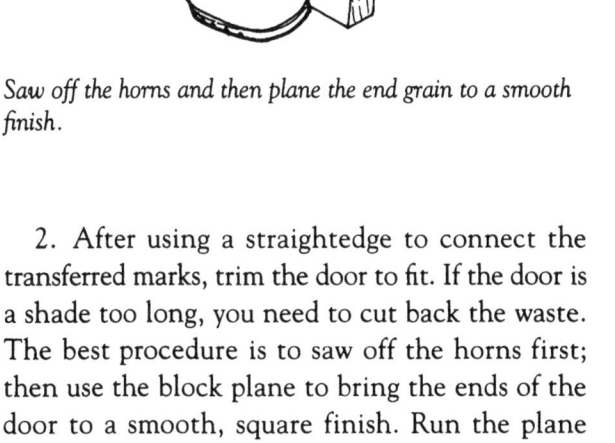

3. Use the smoothing plane to cut back the stiles. Aim for an all-around allowance of between $1/16$ and $1/8$ inch, depending on the desired finish. At this stage, it's best to trim back on the hinge side of the door.

SPECIAL TIP
If the door is too short, the best procedure is to glue and plant a strip between the horns, punch the nails well below the surface, saw off the horns, and then use the block plane to plane back to the mark.

Mortising the Butt Hinges

1. Do a trial fit to make sure that the door fits the jamb. Then set the door hinge-side-up on the floor—so that it is firmly supported by one side of the doorjamb and the two sawhorses—and use the rule and square to mark the position of the hinges (two or more, depending on the size of the door). It's best to have the top of the top hinge about 6 to 7 inches down from the top of the door, and the bottom of the bottom hinge 9 to 10 inches up from the bottom of the door.

2. Decide which way you want the door to open and pencil-mark the hinge stile accordingly; then take the square and the marking knife and mark the length of the hinge on the edge of the stile. Make sure that the knife makes a clean cut across the fibers of the grain.

3. With the two lines in place across the grain, take the marking gauge and lay out two lines with the grain—one for the depth of the mortise, and the other for the back edge of the hinge flap. Place

Hanging and Fitting a Door • 65

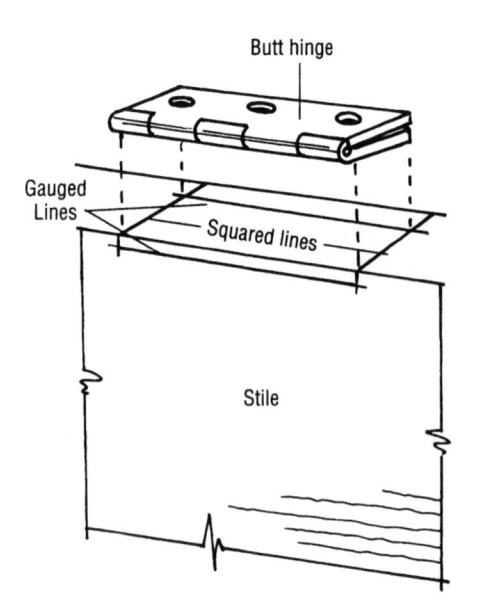

Use the knife and square to mark the length of the hinge, and use the gauge to mark the width and depth of the mortise.

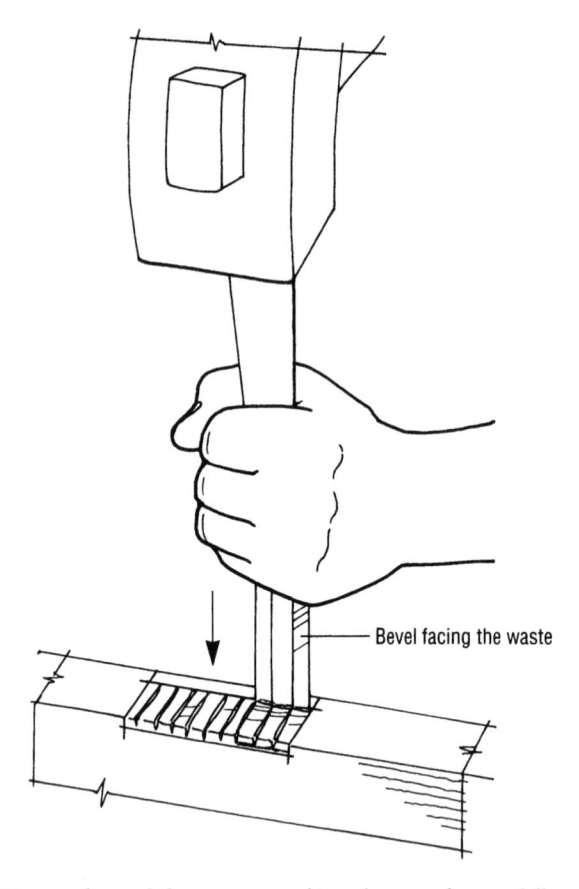

Having chopped the waste in at ¼-inch intervals, carefully cut in all three sides of the mortise.

the back-of-hinge line so that the knuckle of the hinge is set out from the face of the door. Repeat this procedure for both hinge mortises.

> **SPECIAL TIP**
> Note that the distance from the edge of the door to the center of the hinge pin represents half the distance from the face of the door when open to the face of the jamb. If, for example, you want the door to swing so that there is an open joint—with the distance between the face of the door and the jamb being about ½ inch—you will have to place the hinge so that the center of the pin is ¼ inch away from the face.

4. When you are happy with the markings, take the mallet and chisel and run stop-cuts into the wood. Space the cuts about ¼ inch apart and down to within a whisker of the gauged depth line.

5. Hold the chisel vertical so that its back is to the back wall of the mortise—that is, so the bevel is facing the waste—and set the line with a couple of crisp taps of the mallet.

6. Having severed the fibers at ¼-inch intervals and tapped in all three sides of the mortise, hold the chisel horizontally about ½ inch in from one end of the mortise—so that it is bevel up, with the cutting edge on the mortise line—and pare away the waste. If you have defined the mortise with the vertical stop-cuts as described, the waste will come away with three or four well-considered sweeps of the chisel.

7. Finally, clean up the edge of the mortise and test for a good fit. Repeat this whole procedure for the other mortise.

Fixing the Hinges and Hanging the Door

1. Set the hinges in position on the door, drill pilot holes for the screws, and screw the hinges in place so that they are pulled tight and firm in the mortise.

2. Once the hinges are in place, next comes the tricky task of marking out the hinge position on the doorjamb. If anything is going to go wrong, this

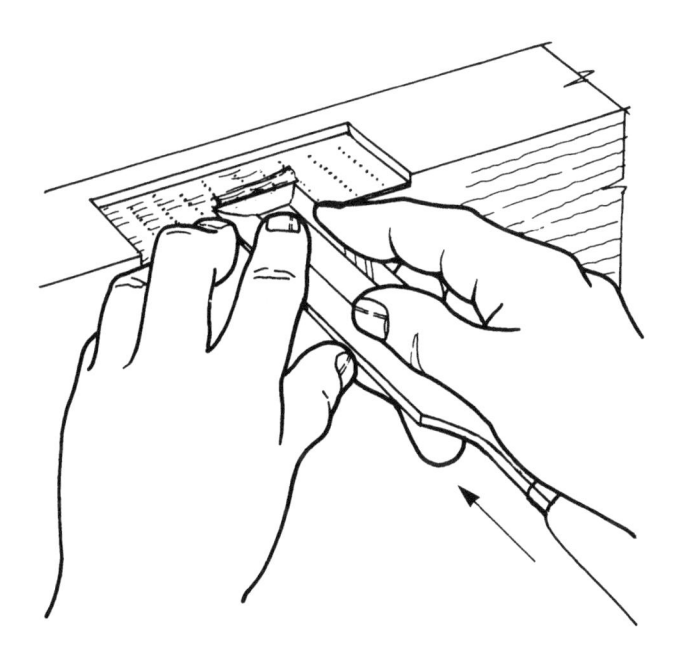

Pare the waste with the chisel—one hand clasping and applying forward pressure, the other hand guiding and steadying the stroke. Pare back to the gauged line.

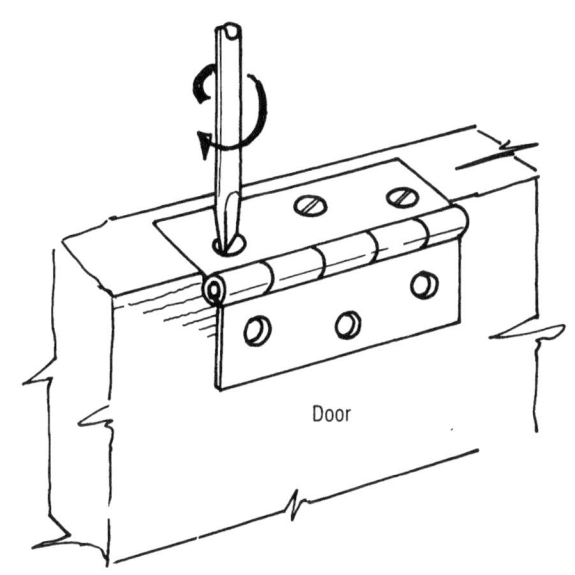

The hinge needs to be a tight fit.

is where it will happen. Set the door into the jamb and against the rabbet, open out the hinge flaps (also called leaves), and use four little wedges to bring the door into position. Have two wedges under the door—one near each stile—and one wedge on each side of the door at a point an inch or so lower than the hinge. Spend time fiddling with the wedges until the door is perfectly placed. When you are happy with the placement, use the knife to mark the end lines on the rabbet.

3. Remove the door and use a pair of dividers to transfer the distance from the back of the door to the edge of the hinge through to the rabbet. Add about 1/8 inch to the measurement to give some clearance—when the door is closed—between the back of the door and the doorstop.

4. When all the guidelines are in place, cut the mortises out in the same way as already described.

5. Set the door in place on the jambs, so that the hinge flaps are in place in the mortises and the door is about half open, and wedge the door up to the proper height.

6. Run a pilot hole in the top hole of the top hinge, and drive the screw home. Try a cautious

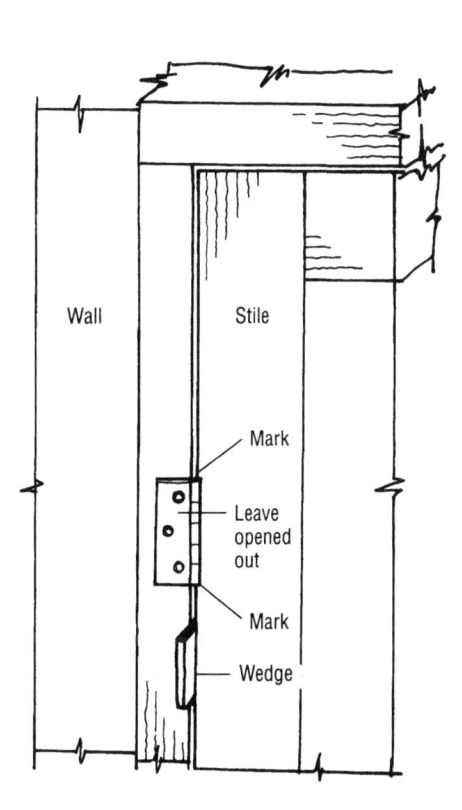

Having wedged the door in place in the frame, use a knife to mark the position of the hinge (the length of the hinge) on the door frame.

Hanging and Fitting a Door • 67

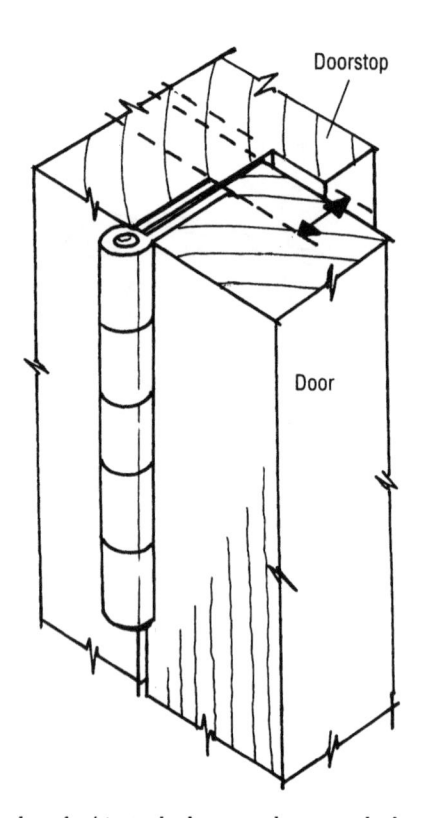

There needs to be 1/16-inch clearance between the back of the door and the stop, to allow for the opening arc and for the buildup of paint.

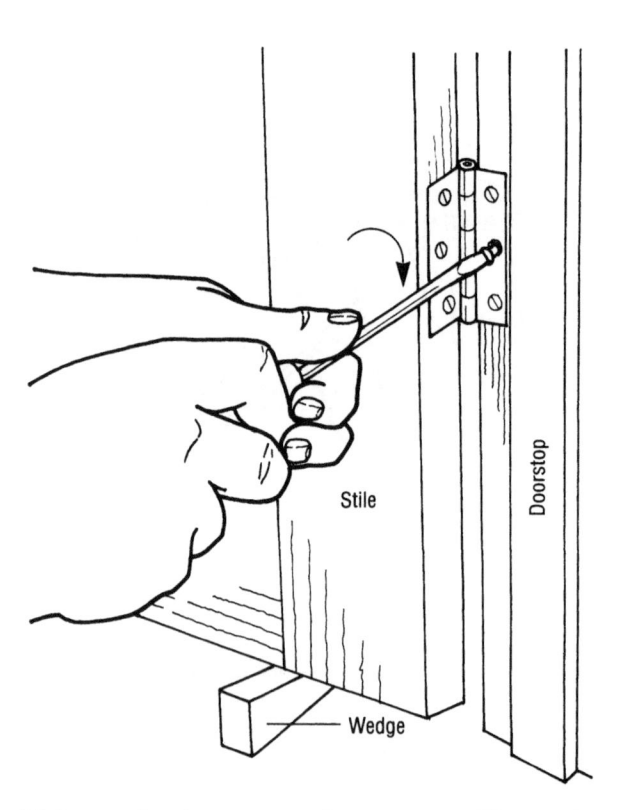

Make sure that the screws are the correct size. They need to be a tight, flush fit.

swing of the door, and if all is correct, drive in the other screws.

Fitting the Mortise Lock

1. Wedge the door open so that you can comfortably approach both sides, and with the lock stile edge-on, take your chosen mortise lock and set it against the face of the door. Center it with the lock rail, and use the body of the lock to establish the position of the key and the spindle holes. Don't forget to allow for the fact that the brass latch plate needs to be recessed so that it is flush with the stile. Run the holes through with suitable-size drill bits.

2. Mark the position of the mortise on the edge of the door, and then use the brace and the Forstner bit to bore out the bulk of the waste. Run a line of overlapping holes along the mortise. When you have cleared most of the waste with the drill, tidy up with a mortise chisel.

> **SPECIAL TIP**
> If you own one of those curious curved chisels known as a "swan neck" lock mortise chisel, now is the time to see if it does the job. The curved neck should enable you to easily scoop out the waste, especially when you need a deep mortise that cuts into the end of the rail tenon.

3. When you have achieved a clean-sided mortise—so that the body of the lock slides neatly into place—push it home, making sure that the main latch or flange plate is hard up against the door. Use a knife and an awl to trace the outline. Use the knife to score across the grain and the awl to run down in line with the grain. If you attempt to use a knife for both tasks, you risk running off course when you mark in the direction of the grain. If you use an awl for both tasks, you will fail to score crisply across the grain.

68 • *Doors*

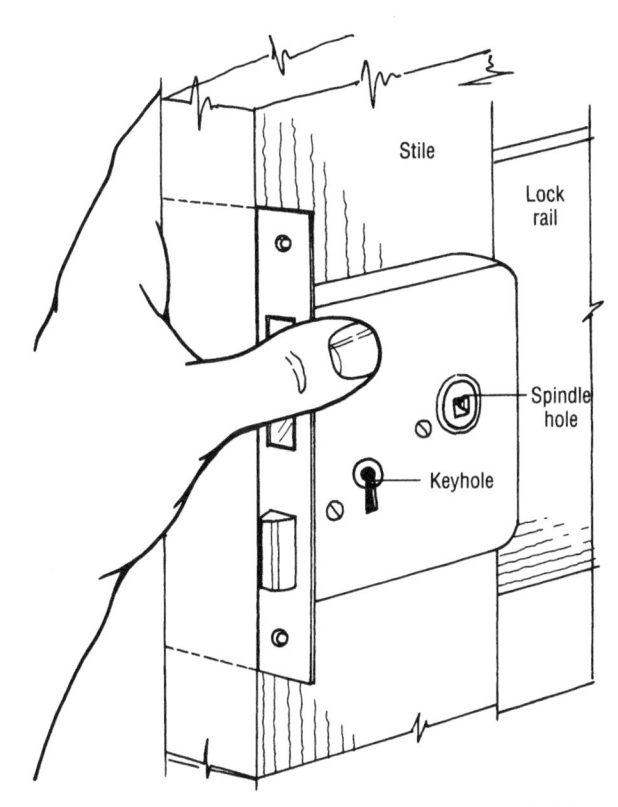

Carefully establish the position of the key and spindle holes.

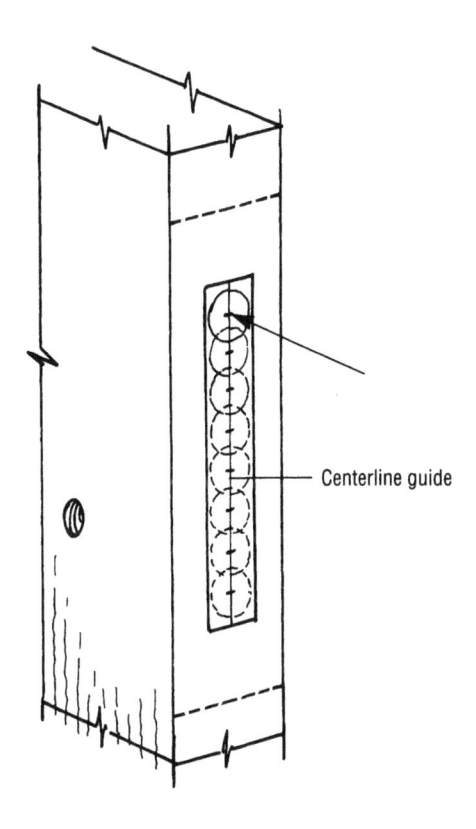

Clear the waste by running a line of overlapping holes. It's best to use the Forstner bit.

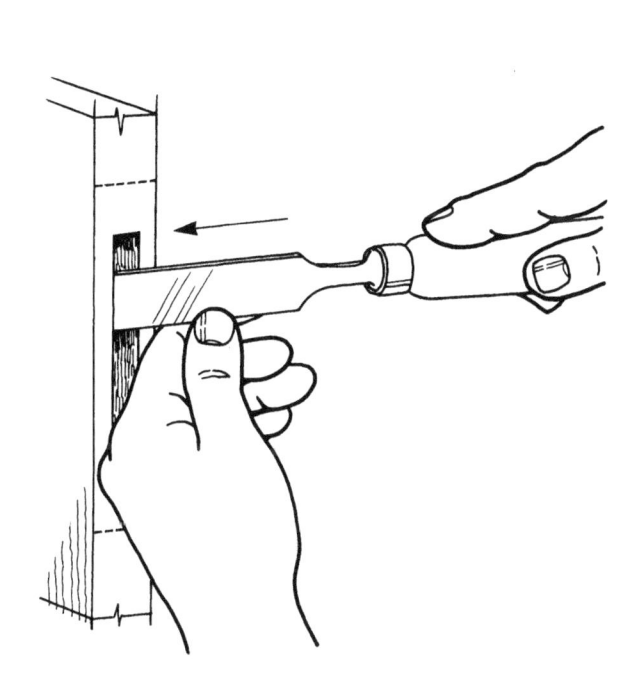

Use the bevel-edged chisel to pare the sides of the mortise to a good fit.

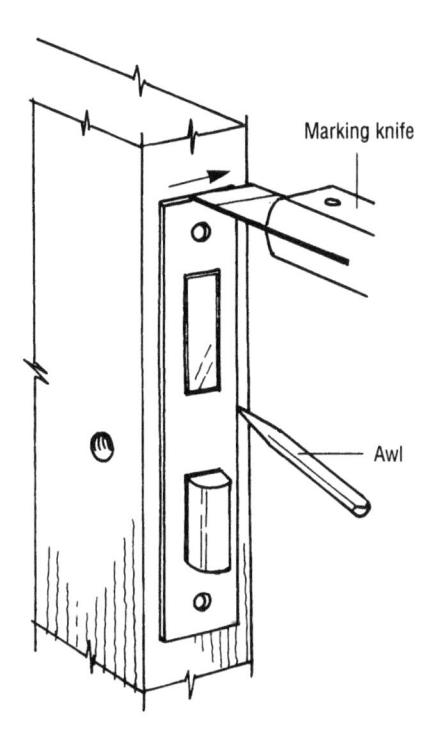

Mark around the flange with a knife and an awl—the knife across the grain, and the awl with the grain.

Hanging and Fitting a Door • 69

4. Chop in the face of the wood at ¼-inch intervals. Then turn the chisel over so that the bevel side is facing the wood and pare from center to top and then from center to bottom. Take it nice and easy, until the latch plate is a clean fit. Brush away the debris, slide the body of the lock into the mortise, and fix the latch plate in place with the screws provided.

5. Close the door three-quarters of the way, and use the knife to transfer the position of the latch through to the doorjamb. Finally, score around the striking plate, cut the mortise for the latch, pare out the recess for the striking plate, and screw the whole works home in much the same way as already described.

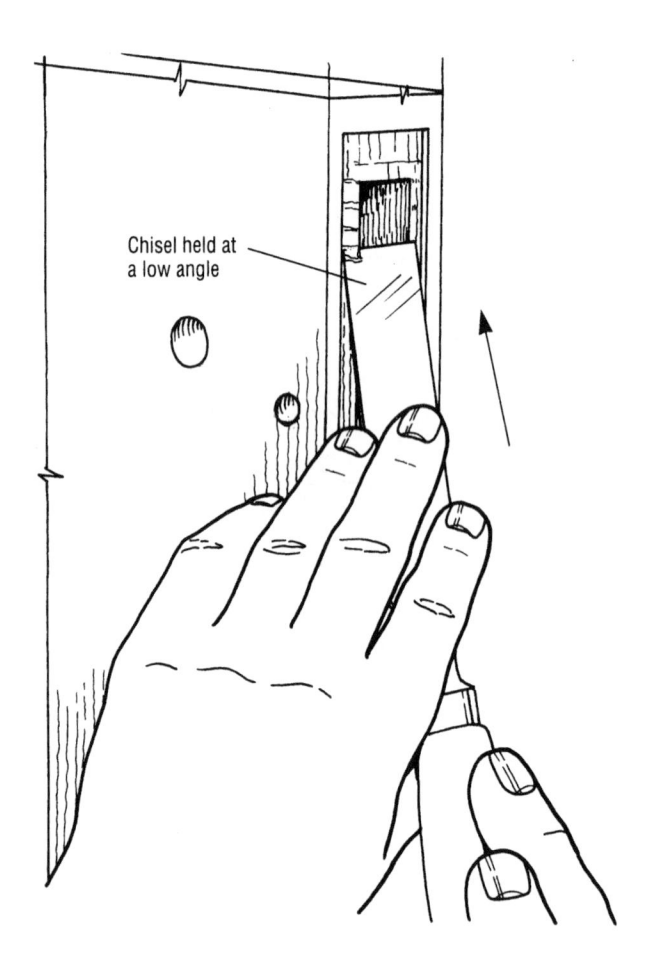

Hold the chisel at a low angle and pare back the waste. Work with the flat back of the chisel uppermost, and cut from the center to the end.

SPECIAL TIP
A good way of pinpointing the position for the striking plate is to smear a little dark oil from your honing stone onto the end of the latch or bolt and let it strike home, transferring oil through the jamb. If you don't like this method, then open the latch and slowly close the door so that the latch marks the edge of the doorjamb.

Fixing the Doorknobs

1. Having made sure that the door swings neatly on its hinges and the latch clicks smoothly home, next comes the task of fitting the knobs, rose plates, escutcheons, and spindle. You have to be mindful at this point that there are two types of knobs and fittings: knobs that are separate from the rose plates, and knobs that are an integral part of the rose plate. Both types have their good and bad points. The separate knobs sometimes causes problems because the little grub screw easily falls out, but the rose is relatively easy to screw to the door. Although the integrated option manages without the grub screw, the design is such that it is sometimes difficult to screw the rose to the door because the knob gets in the way of the screwdriver. The best advice is to choose the knob you like best and take it from there. Here we're using the old-fashioned option, with the rose plates and the handles being separate components.

2. When you have obtained all the parts, have a trial fit to see how it all comes together. Screw one knob to the end of the spindle, slide the spindle through the one rose plate and on through the lock, set the other rose plate on the spindle and hard up against the door, and then set the other knob on the spindle. If you've got it right, the spacing will be firm, but not so firm that everything is tight.

3. When you are happy with the fit, slide the spindle in place and screw on the two rose plates. Have the screw holes set left and right, and be careful that you don't scratch the door or the plates.

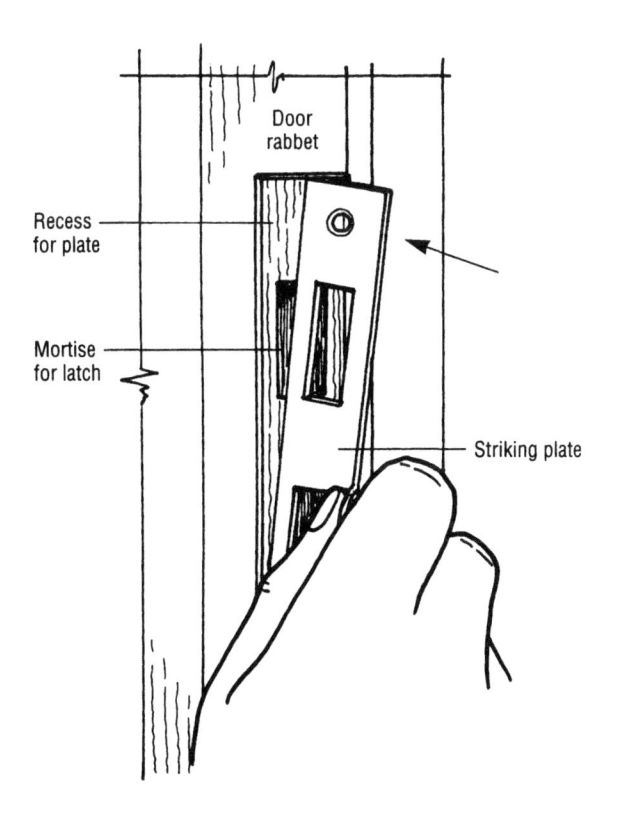

It's important that the striking plate fit flush.

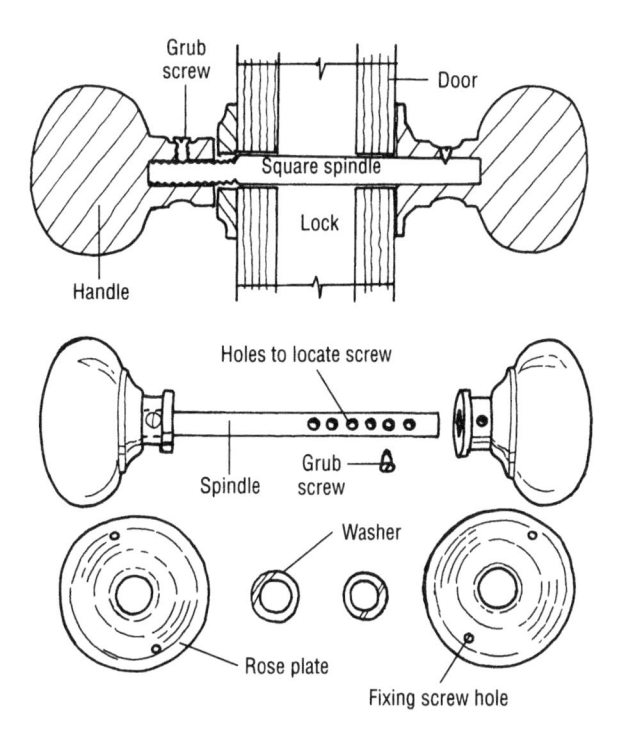

Doorknob types.
Top: *A knob with one side having an integral rose plate.*
Bottom: *A knob with separate rose plates.*

4. Screw one knob on the spindle and arrange the fixing screw (the little grub screw) so that it is on the top side of the knob. Slide the other knob on the spindle so that the fixing screw hole is at the top, and screw it in place.

5. Finally, fit the escutcheon plates over the keyhole, and the job is done.

Afterthoughts and Options

• Don't be tempted to use pressed steel hinges. They are useless.

• Don't be tempted to reuse old screws or make do with screws that are a little bit too big or too small. If you are going to all the trouble of hanging a door, it's a bad idea to lose the battle for the sake of a screw.

• The order of painting a paneled door is as follows: moldings, panels, rails, and stiles. If possible, paint the door before you hang it.

• When using a power drill to bore out the spindle hole and the like, be careful not to leave a mess of splinters.

Hanging and Fitting a Door • 71

Door Designs

DESIGN 1

If you are looking for a door with a history, a door that is going to keep the foe at bay, a door that is going to put your woodworking skills to the test, this is the one for you. The design is both a delight and a challenge. When we first saw this door, our immediate reaction was, how the heck did those old English woodworkers manage to build such a beauty—all that bent wood, and those forged iron roves? As far as the making is concerned, the huge oak planks with the edge rabbet are easy enough, but the actual putting together is a huge challenge. Our advice is to start out by making contact with a friendly blacksmith. The design is inspired by an ancient four-plank door belonging to St. Botolph's Church in Hadstock, Essex, England. The wide planks are rabbeted and held together by D-section oak, with the D-sections being clinched to the planks with iron roves. The rather curious technique of fixing the D-sections with roves almost certainly had its beginnings in boatbuilding.

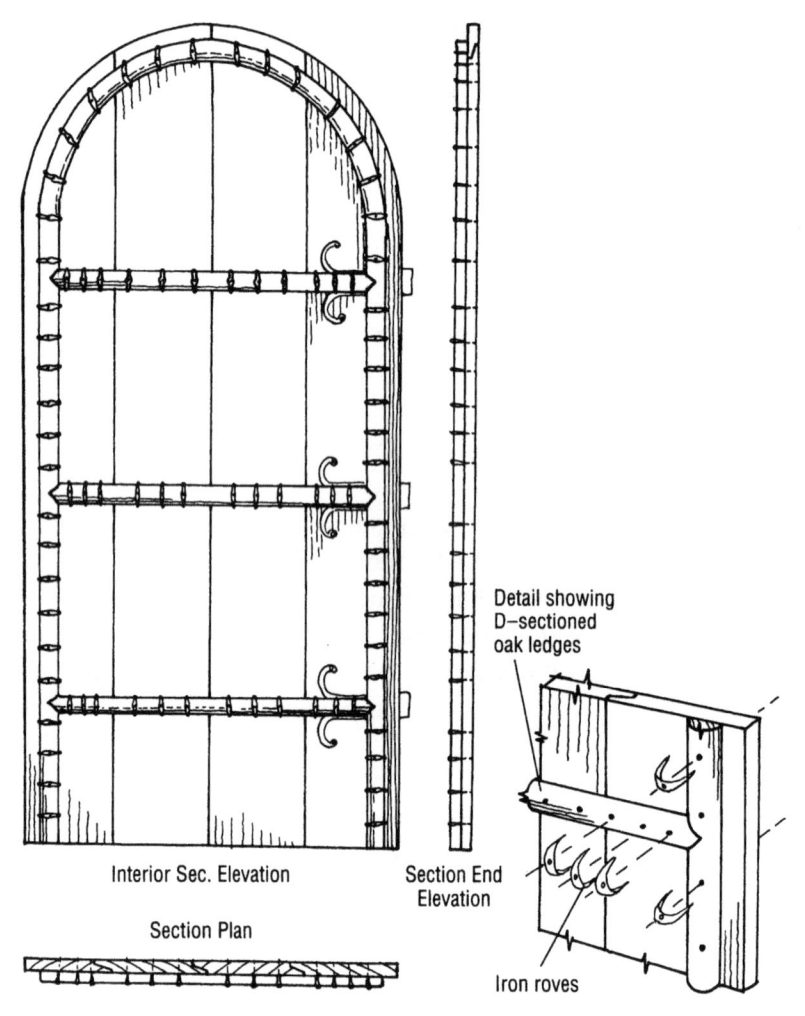

Interior Sec. Elevation

Section Plan

Section End Elevation

Detail showing D-sectioned oak ledges

Iron roves

Door type and location: *Basic slab plank door; good for a traditional setting such as a church, barn, or cottage.*
Construction and joints: *Slab plank construction with rabbeted joints, D-sections, and iron rove reinforcements.*
Skill level: *A difficult door to build, because of the need to steam and bend the D-sections.*

DESIGN 2

Not only is this a door with a long history, but the construction techniques are a pure delight. Though you might say that this is a plain, simple, no-nonsense door—after all, what could be simpler than the rabbeted edges and the ledges?—the design is very sophisticated. This door would look great in a barn-type house or even an old farmhouse. The five planks are rabbeted, with the three ledges being housed on opposite faces of the door. The unusual shape of the ledges and the pegs make for a very solid and stable construction.

Door type and location: *Basic slab plank door; good for a traditional setting such as a church, barn, cottage, or log cabin.*

Construction and joints: *Plank construction, with rabbeted joints and housed dovetail ledges to one face.*

Skill level: *An easy door to build, though you must make sure that the dovetail ledges are a tight fit.*

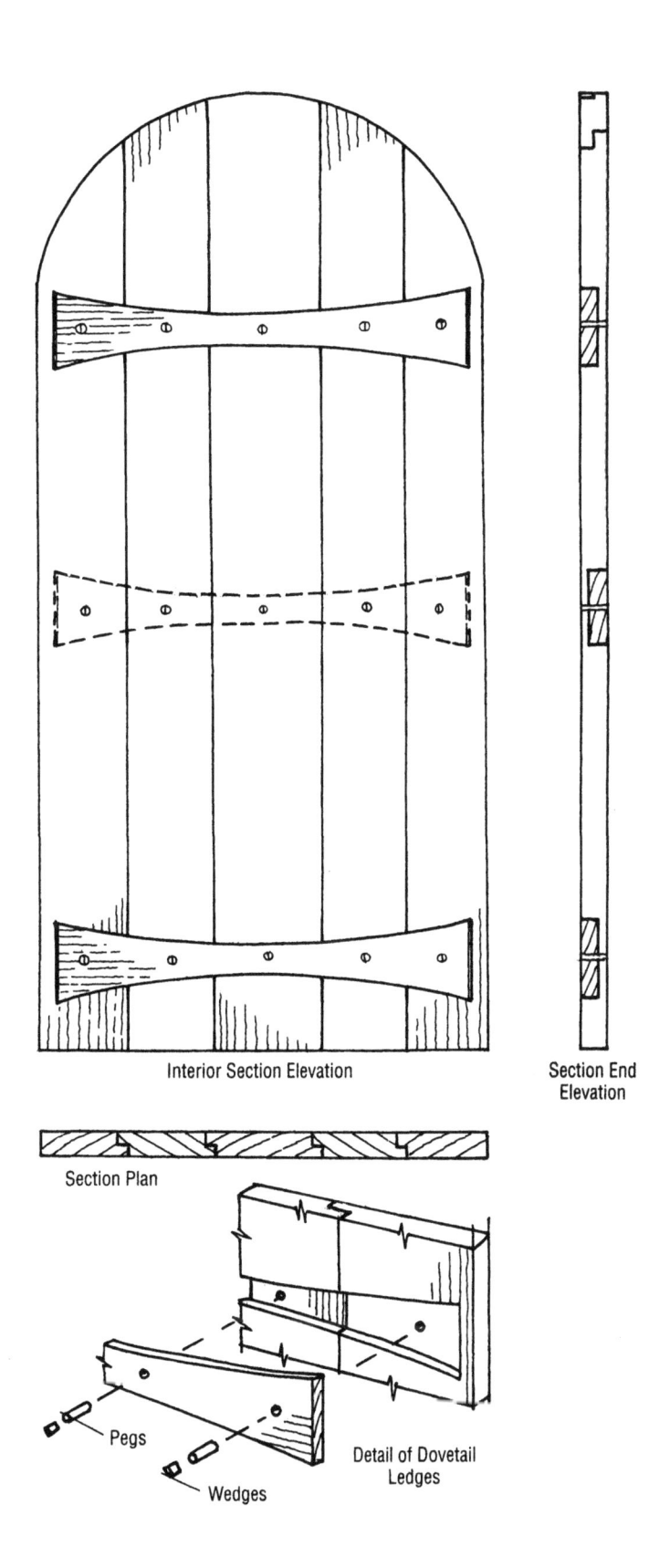

Interior Section Elevation

Section End Elevation

Section Plan

Pegs

Wedges

Detail of Dovetail Ledges

Door Designs • 73

DESIGN 3

This beautiful door draws its inspiration from the west door of Kempley Church in Gloucester, England. The original door is almost certainly from the twelfth century. The door consists of three planks that are counter-rabbeted, with the whole works being held together by loose tenons. The massive door is also held together with the most exquisite forged iron strap hinges. If you have a mind to build this door, first sit down and consider all the implications of building such a masterpiece. Are your skills up to it? Also, you'll have to find a blacksmith who's willing and able to make the strap hinges.

Door type and location: *A basic slab plank door; good for a traditional setting such as a church, barn, or cottage, but would also look good in a contemporary setting.*

Construction and joints: *Plank construction with counter-rabbets and loose tenons. The door is reinforced with forged ironwork.*

Skill level: *A very difficult door to build. You have to use massive sections to achieve the locking planks.*

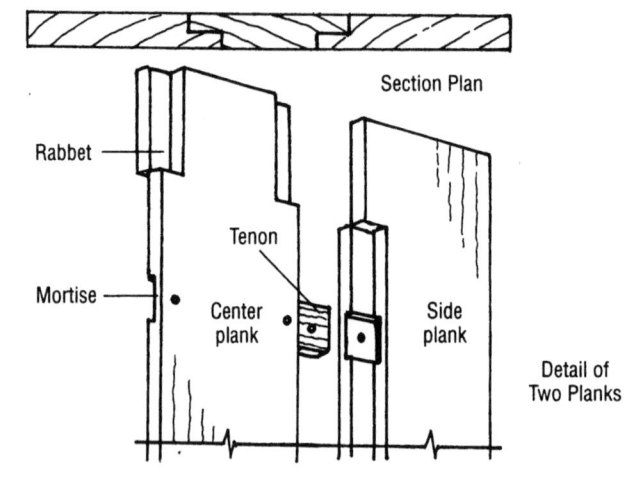

Exterior Section Elevation

Section End Elevation

Section Plan

Rabbet

Mortise

Tenon

Center plank

Side plank

Detail of Two Planks

DESIGN 4

This door draws its inspiration from a twelfth-century door belonging to St. Alban's Cathedral, England. Other than the wonderful iron hinges, not much of the door remains. If you get half a dozen oak planks, set to work building a straightforward ledged-type door, and then make contact with a talented blacksmith, at the end, you will have an amazing item. Once again, the secret of this particular door is to start by finding the blacksmith. You could say that the success of this door hinges on the hinges.

Door type and location: *Slab plank door; good for a traditional barn, church, or cottage setting.*
Construction and joints: *Butt-edged board, ledges to the back; the door is reinforced with nail-clinched ironwork.*
Skill level: *The woodwork is easy, but the ironwork requires great skill.*

DESIGN 5

This classic ledged door draws its inspiration from the thirteenth-century south door of St. Laurence and All Saints Church in Eastwood, Essex, England. The three oak planks are rabbeted and held together with housed dovetails. Certainly a door of this character would look wonderful in a barn-type house, but the design is so simple and direct that it would also work in a contemporary house. If you want to build a door in the classic Old English tradition, perhaps this is the one.

Door type and location: *A slab plank door; good for a traditional setting such as a church, barn, or cottage, but would look equally good in a sophisticated modern building.*
Construction and joints: *Plank construction, with rabbeted joints and sliding wedge-dovetailed ledges.*
Skill level: *A tricky door to build. You would have to perfect your sliding dovetail techniques by making several practice runs on scrap wood. It's best to build a full-size mock-up.*

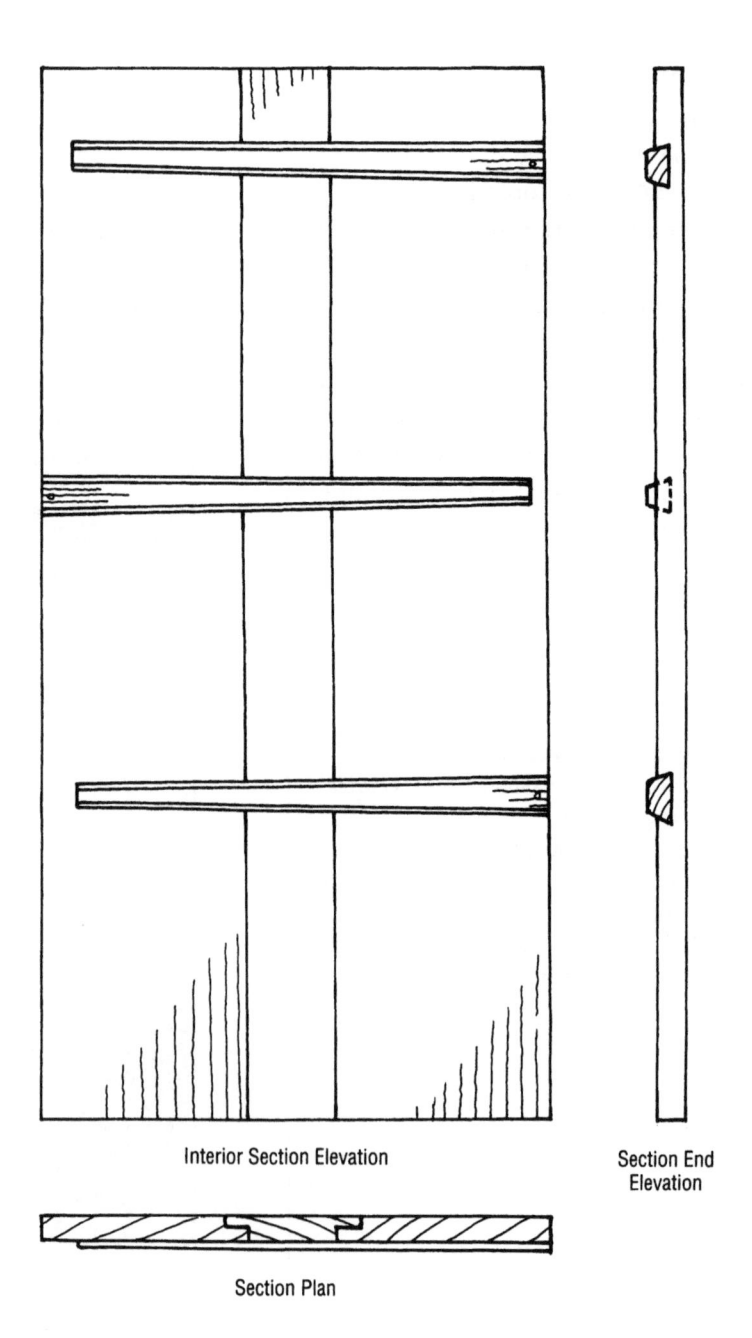

Interior Section Elevation

Section End Elevation

Section Plan

76 • *Doors*

DESIGN 6

This design was inspired by the south door of the Church of All Saints in High Roding, Essex, England. When we first saw this door, we were somewhat amazed by the complex construction. The whole structure is cross-framed and then paneled, with the small panels sliding in V-section grooves within the framing. But more than that, the rails are dovetail lapped into the stiles. Our first impression was that it's as if the maker had been trying to create a puzzle. Perhaps even more amazing is that the run of the grain within the stiles suggests that the carpenter searched out naturally curved pieces of wood to fit the door. If you enjoy a challenge and have a yen to cut countless laps, dovetails, and grooves, this is the door for you.

Door type and location: *Slab plank door with a lancet head; good for a traditional setting such as a church, a barn, or cottage or in an Arts and Crafts setting.*
Construction and joints: *Complex framed and paneled construction, with a grooved frame, stub tenons, bare-faced lap dovetails, half laps, and pegs.*
Skill level: *A tricky door to build because there are so many close-fitting parts. You also have to decide up front how you are going to construct the curved head: Are you going to search around for a piece of wood to fit the curve? Are you going to make it up in sections and joint it with loose hammerhead wedges? Or are you going to try something daring like steam-bending or laminating?*

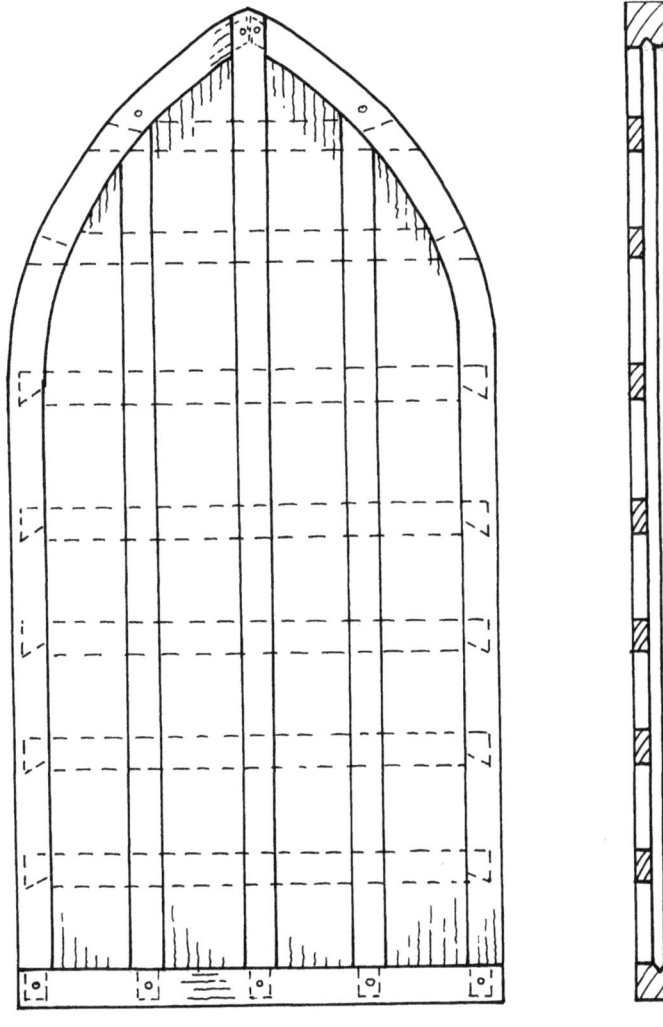

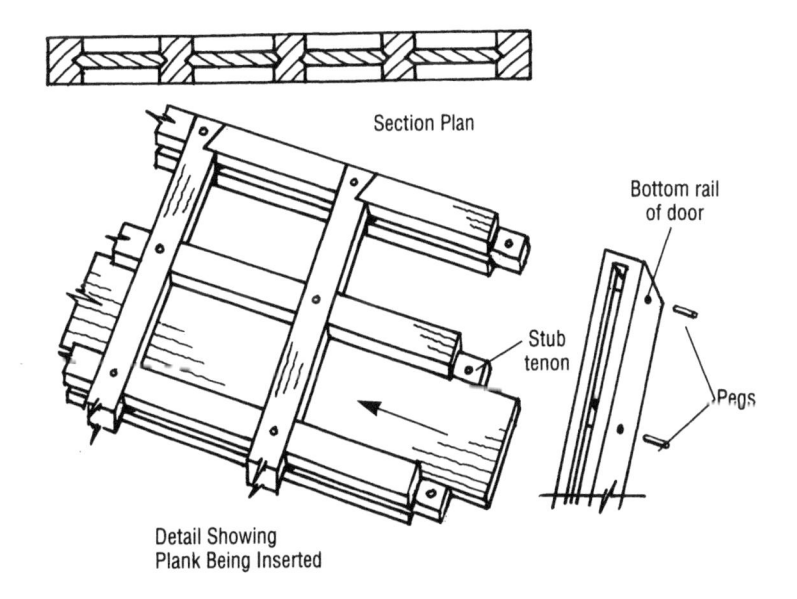

Exterior Section Elevation

Section End Elevation

Section Plan

Detail Showing Plank Being Inserted

Bottom rail of door

Stub tenon

Pegs

Door Designs • 77

DESIGN 7

The thirteenth-century door of Dore Abbey, Hereford, England, inspired this door. It is made from five massive planks with an equally massive trefoiled head, all with chamfered ledges to fit. The whole works is held together with oak pegs. Although it's top-heavy and cumbersome, it's a good choice if you're looking for a door with a strong Gothic feel.

Door type and location: *Basic slab plank door; good for a traditional setting such as a church, barn, or cottage. It might also look good in a Victorian Gothic house.*

Construction and joints: *Plank construction with V-grooves and tongues, with a fretted and applied headboard. It's all held together with nails clinched through to the ledges.*

Skill level: *An easy door to build—very basic.*

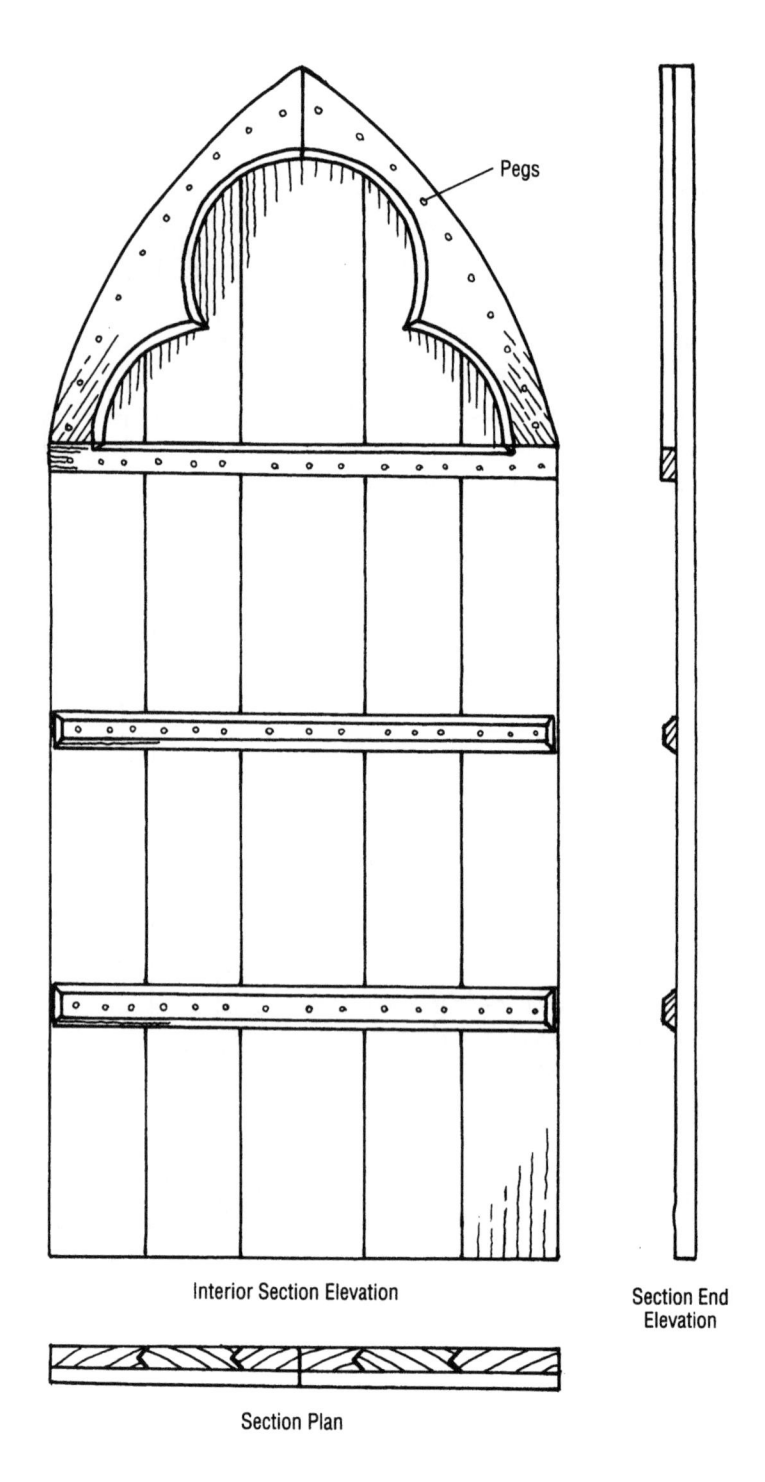

Pegs

Interior Section Elevation

Section Plan

Section End Elevation

DESIGN 8

This design draws its inspiration from the north door of the Church of St. Michael in Fobbing, Essex, England. This fifteenth-century door is odd in that the jointing techniques are so unusual. It's almost as if the maker had been looking for the most awkward way of putting a door together. For example, the planks are V-edged and shunted together so that the surface is rippled, and the massive strap hinges are fitted to the rippled face in such a way that they make contact only with the peaks. When we saw this door, our first thought was, "What a nightmare!" Our second thought was that the whole thing would be immensely improved by having the hinges fitted on the level side of the door, and our third thought was that the hinges are beautiful—almost modern in design.

Door type and location: *A basic slab plank door; good for either a traditional or a modern setting.*

Construction and joints: *Plank construction with grooves and tongues; the faces of the boards are molded, and the whole thing is held together with massive forged iron strap hinges. The ironwork is vital to the construction.*

Skill level: *An easy door to build, but you'll need to find a talented blacksmith.*

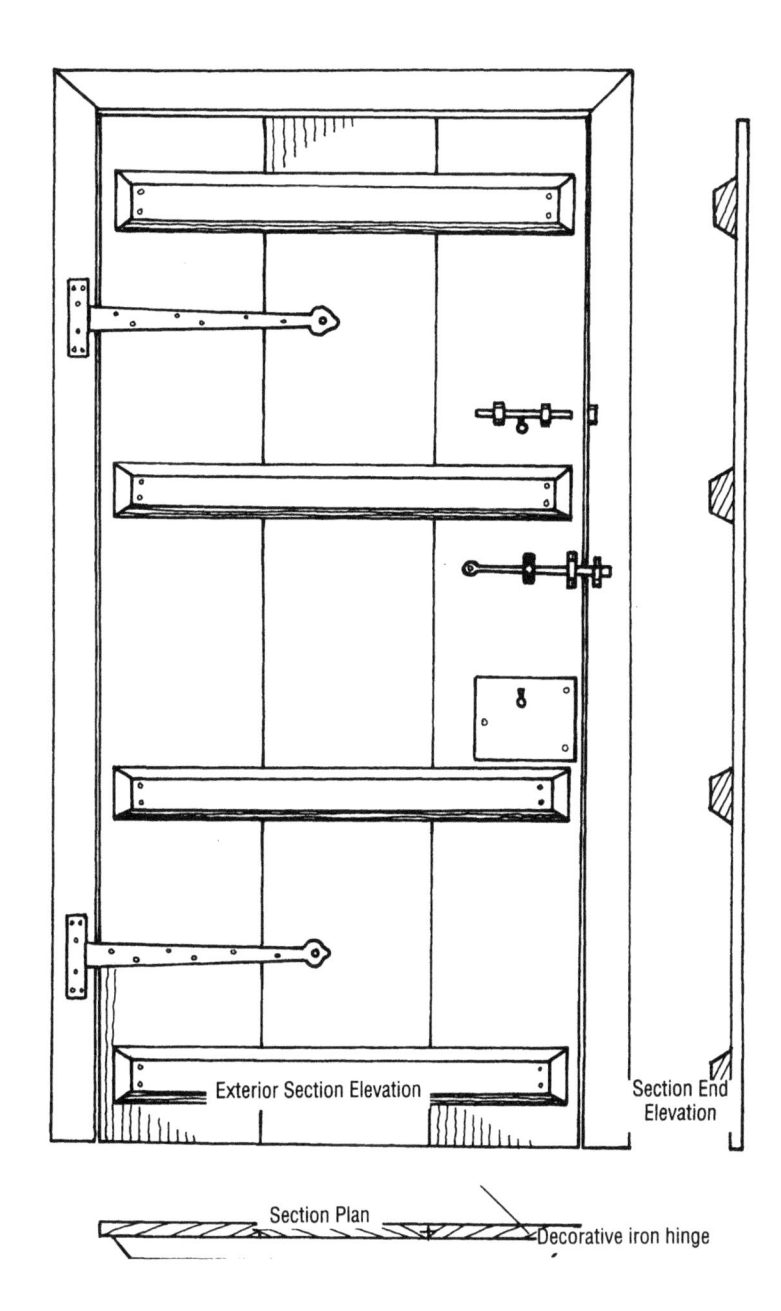

Exterior Section Elevation

Section End Elevation

Section Plan

Decorative iron hinge

Door Designs • 79

DESIGN 9

An attic door found in the old St. George's Inn in Southwark, London, England, inspired this door. We can't be certain about the date, but our research suggests that it was made in the last quarter of the seventeenth century. The three planks are tongued and grooved, and the whole structure is held together with the massive ledges. It's amazing that a door like this has been opened and closed hundreds of thousands of times over the last 300 years or so, yet there it is, still swinging and doing its job. If you live in an old wooden house or are going to refurbish a barn house, a classic door like this would be perfect. It must be made in large-section wood, however.

Door type and location: *A traditional plank and ledge door; good for a traditional setting such as a farmhouse or cottage.*
Construction and joints: *The massive planks are grooved and fitted together with loose tongues. The ledges hold the planks together.*
Skill level: *An easy door to build.*

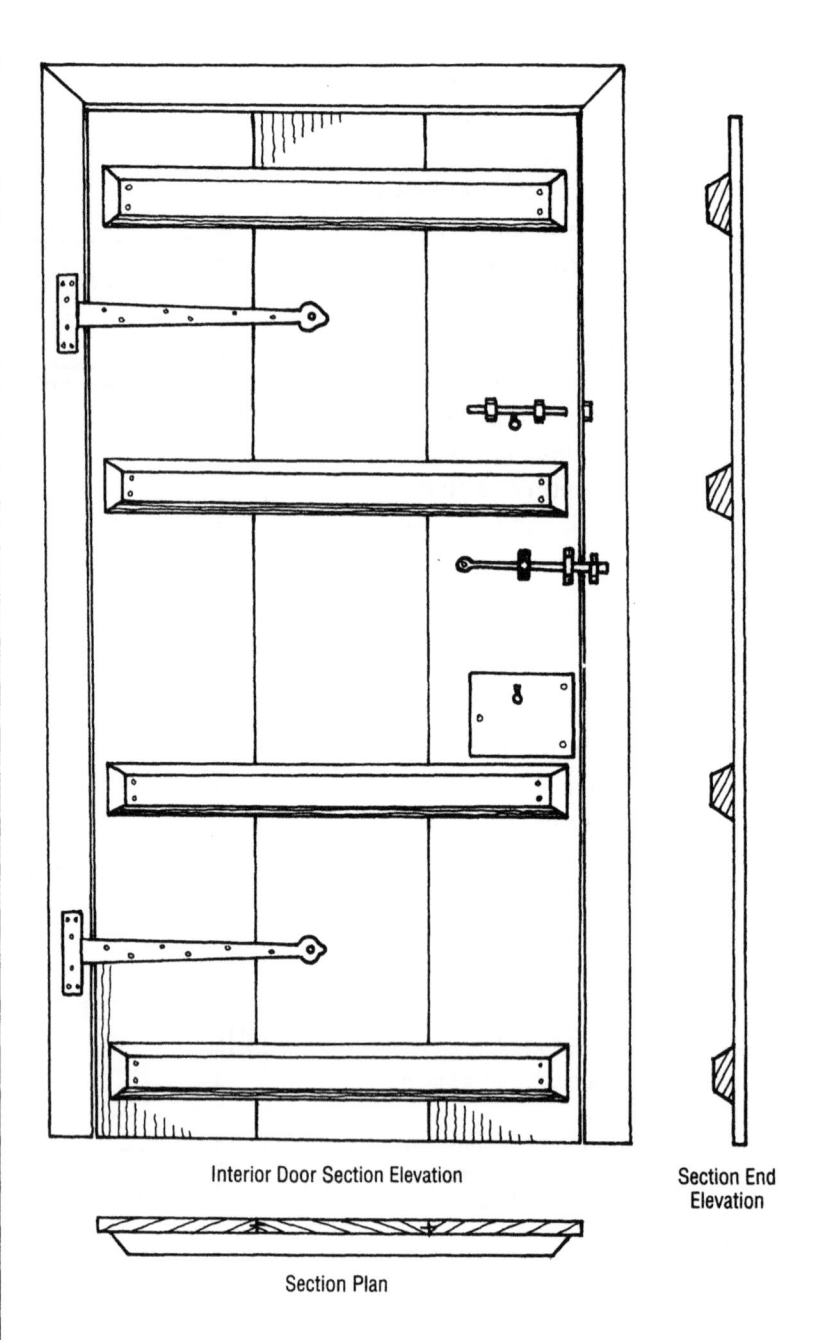

Interior Door Section Elevation

Section End Elevation

Section Plan

80 • *Doors*

DESIGN 10

If you live in a grand house, you may want to build pediments over the doors. It's best to use an easy-to-work wood such as lime or basswood. Work the moldings with a plane or a router, cut the curved profiles on a cross saw, and then carve the details.

Door type and location: *Pediments designed to fit over interior doorways; good for an eighteenth- or nineteenth-century house.*
Construction and joints: *Built up from a mix of fretted profiles and moldings and held together with glue and dowels. Filled with plaster and then painted.*
Skill level: *Very easy to make—no complex jointing needed.*

Door Designs • 81

DESIGN 11

It might look as if these two doors were built centuries apart, but they were both made in eighteenth century. To be more precise, they were both made for the same Pennsylvania German house in 1783. It's strange that although the two doors were made in the same year and perhaps even by the same hand, one is so naive and the other is so sophisticated. Were the doors a whim of the maker or of the owner? Or was there a tradition in Germany of harking back for less important doors, and being up-to-date for prestigious doors?

The traditional batten and ledged door in the illustration has an opening transom window. The door is fitted with massive cross-garnet hinges and a security bar. The classic framed and paneled door has a pediment above and denticulate details on the cornice.

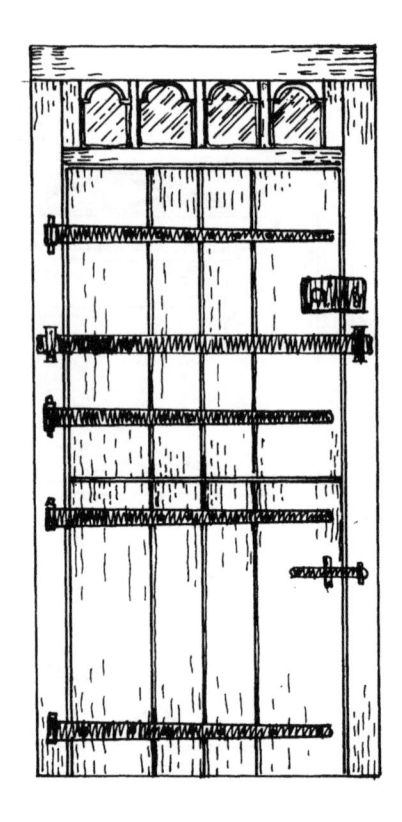

Left:
Door type and location: *A traditional batten and ledged door; good for a farmhouse or cottage; best as a door leading through to the garden.*
Construction and joints: *The battens are butted together and held with ledges, and nails are run through the strap hinges and clinched over on the back of the ledges. There is a small amount of jointing above the transom—mortise and tenons and stub tenons.*
Skill level: *The door is easy, but the transom glazing requires the ability to cut mortises and tenons and rabbets.*

Right:
Door type and location: *A framed and paneled interior door designed for a prestigious room; this would look good in an eighteenth-century house.*
Construction and joints: *Traditional framed and panel construction, with the full range of mortises and tenon joints, grooving, and some basic carving.*
Skill level: *A fairly easy door to build. Much depends on your choice of wood; it's best to use a straight-grained type such as cherry.*

DESIGN 12

If you are building a timber-framed house and are searching for an attractive front door, this may be the one. This pretty eighteenth-century door has a glazed transom above, a bracketed porch head, and shutters to the sides. The charm of this design is the understated, low-key feel of the woodwork—it's almost as if the maker went out of his way to keep the details simple and plain. The headpiece is leaded, the door itself is painted brick red, and all the fittings are brass.

Door type and location: *A traditional framed and paneled door; good for the front door to a wood framed cottage.*
Construction and joints: *Traditional framed and paneled construction with the full range of mortise and tenon joints. The frame is grooved. The glazed transom requires mortises and tenons and stub tenons. The side brackets are fretted and built up.*
Skill level: *Middling to easy—no more than a basic door with lots of fancy curlicues.*

Door Designs • 83

DESIGN 13

This group of doors was used in a development of Queen Anne–style cottages built in New York in 1881. The wonderful thing about simple framed and paneled doors of this character is that there are countless ways of arranging the panels.

Door type and location: *Basic framed and paneled doors, designed as porch doors; good for a framed cottage.*
Construction and joints: *Traditional framed and paneled construction, with the full range of mortise and tenon joints.*
Skill level: *Though the doors are pretty basic, the small panels mean that your measuring must be dead-on perfect.*

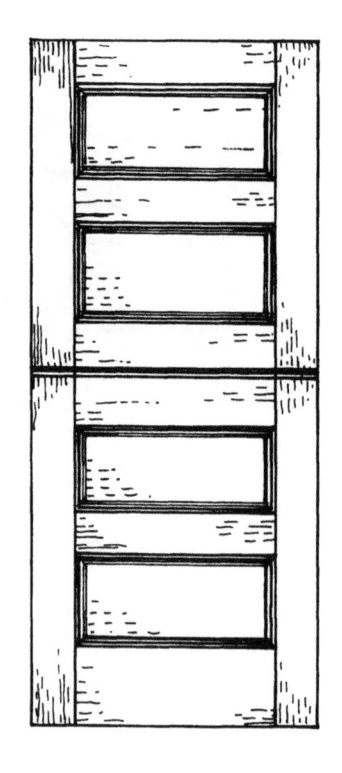

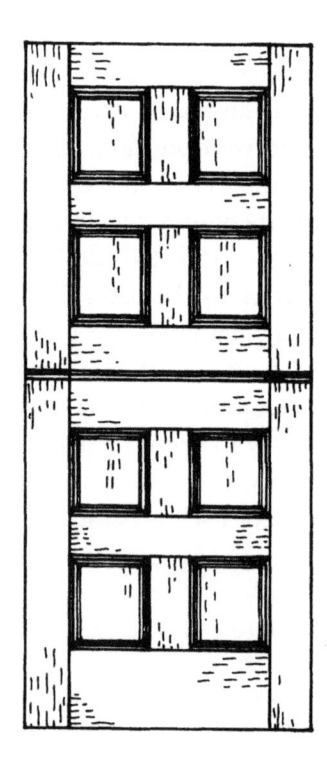

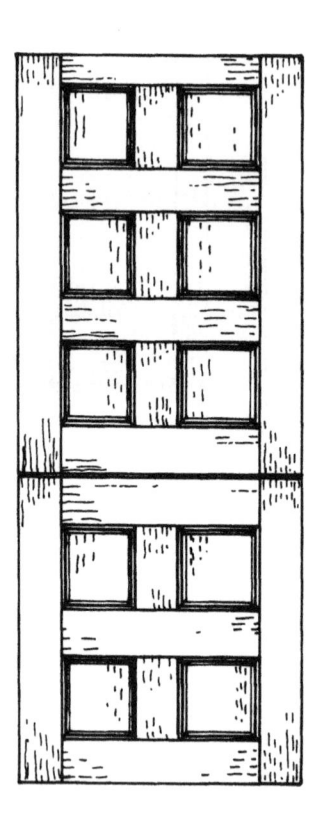

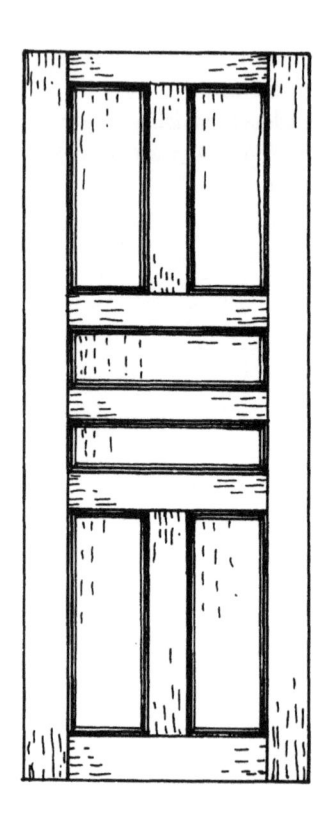

84 • *Doors*

DESIGN 14

For many door makers, the Victorian period was the tops; the doors were exciting, dynamic, and innovative. Of course, some of the doors were undoubtedly gross and out of proportion, but it's plain to see that the door makers were at least prepared to try something new.

Top left: Victorian framed and paneled door with raised panels, bolection moldings, and an oval glazed panel above.

Top right: Victorian framed and paneled main door with six panels. Two panels are glazed, and there is generous bolection molding on the four bottom panels.

Bottom left: Victorian framed and paneled door with an unusually wide locking rail. The top panel has generous bolection molding.

Bottom right: Late Victorian framed and paneled door with two glazed panels above.

Door type and location: Traditional framed and paneled doors; best for Victorian and Edwardian houses.
Construction and joints: Traditional framed and paneled construction uses the full range of mortise and tenon joints. The bolection moldings are mitered and nailed to the frame.
Skill level: Overall, the construction is simple, but the mitering needs to be done with great care. The oval window is tricky because you need to build a curved molding to hold the glass.

Door Designs • 85

DESIGN 15

If you are searching for a classic front door design—a design with raised panels and some degree of glazing—one of these doors might hit the spot. Designed in 1881 for a New York timber framed house, these doors are formal without being heavy and overbearing.

Top: A five-panel door with glazed lights above, raised panels, generous bolection moldings throughout, a wide architrave surround, and planted plates to the corners.

Bottom: An eight-panel door with glazed panels above, bolection moldings throughout, raised panels, and plates to the corners of the architrave. The arrangement of the small panels gives the design a feeling of strength and security.

Door type and location: Basic framed and paneled doors; best in a late-nineteenth-century framed cottage. These doors were designed variously to serve principal rooms and to lead to porches.
Construction and joints: Traditional framed and paneled doors with glazing above, the full range of mortise and tenon joints, plus various stub tenons for the glazing bars.
Skill level: Fairly difficult because of the glazed transom above. The corner plates on the architrave eliminate the need to cut miters.

86 • Doors

DESIGN 16

This rather serious and squat front door, with its Queen Anne–style details, was designed and built in 1880 for a London house. Although it certainly has all that you would expect of a door of this period—panels, glazing, molding, brass fittings, and all the rest—it is rather melancholy.

Door type and location: *A late-eighteenth-century framed and paneled door; best in the correct period house.*
Construction and joints: *Traditional framed and paneled construction, with the full range of mortises and tenons. The glazing bars are stub-tenoned.*
Skill level: *Difficult because the massive construction calls for lots of effort, and the glazing bars call for ultra-precise cutting.*

DESIGN 17

This Victorian Gothic sash door, built for an estate cottage, is both romantic and practical. The lancet shape speaks of silk top hats, crinolines, vicarage tea parties, and roses around the door, and the glazed panels are designed to let in the maximum light. If you are looking to build a period door for your orangery, or maybe a door to open onto a south-facing garden, this could be the one. The glazing bars are V-tenoned into the frame, and the iron studs are made from large, hand-forged, rose-head nails.

Door type and location: *Framed and ledged; best in a Victorian Gothic setting or as a garden door.*
Construction and joints: *Framed and ledged, with the frame being built up. The rails are mortised and tenoned; the glazing bars are shaped and variously lapped and stub-tenoned. The infill in the lower half is clinched through to the ledges.*
Skill level: *A very difficult door to make because of the need to build up the curves and to fit all the glazing bars.*

DESIGN 18

These doors—designed specifically for where there is a need for additional natural light—are wonderfully fitting for their purpose. They are small and modest in size, and yet at the same time they are delicate and decorative. If you live in a small traditional town house and are looking for an inner door—one that divides the hall off from the back kitchen or perhaps a small front door—then one of these doors would be a good choice. If you are using salvaged doors, the moldings and the lead work need to be sound.

Door type and location: *Traditional framed and paneled front doors; good for cottages, but the doors should be protected by a porch.*
Construction and joints: *Traditional framed and paneled, with the full range of mortises and tenons. The leaded lights must be made by a specialist, or you could opt to use salvaged windows.*
Skill level: *Although the doors are pretty straightforward to build, the glazing requires extra effort. The door at the top left is difficult because the slender D-section glazing bars have to be curved and then mitered together.*

Door Designs • 89

DESIGN 19

These wonderfully elegant double entrance doors, built in 1881, would be perfect as vestibule doors. With their classic framed and paneled structure and the leaded stained glass above, they are both practical and welcoming. If you are looking for a pair of high doors to divide up a large entrance hall, these might fit the bill.

Door type and location: Traditional framed and paneled doors with decorative glass; good for vestibule doors or doors leading through to a covered porch.
Construction and joints: Framed and paneled with the full range of mortise and tenon joints, with machine-worked panels at bottom.
Skill level: Straightforward construction, but a specialist would need to be brought in for the glazing.

DESIGN 20

These double sash doors, designed for a lakeside cottage in Boston, Massachusetts, in 1890, are both elegant and practical. They provide a good strong barrier, yet they allow the light to shine through. Before beginning to build these doors, get details such as the leaded lights and the carved panels from an architectural salvage specialist. Or you could commission a stained-glass artist to make windows of your own design.

Top: Framed and paneled with etched and stained glass above, heavy moldings throughout, and turned bosses on the rail.

Bottom: Framed and paneled with leaded glazing above; the bottom panels are raised and heavily carved.

Door type and location: Basic framed and paneled doors; good for a grand house setting, as vestibule doors or perhaps as doors to a covered porch.
Construction and joints: Framed and paneled, with the full range of mortises and tenons. Both sets of doors present problems when it comes to making the bottom panels—one requires lots of carefully mitered moldings, plus the turned bosses, and the other requires that the panels be carved.
Skill level: Although the door construction is pretty straightforward, the glazing is an added problem. The carved panels are easy if you enjoy relief carving.

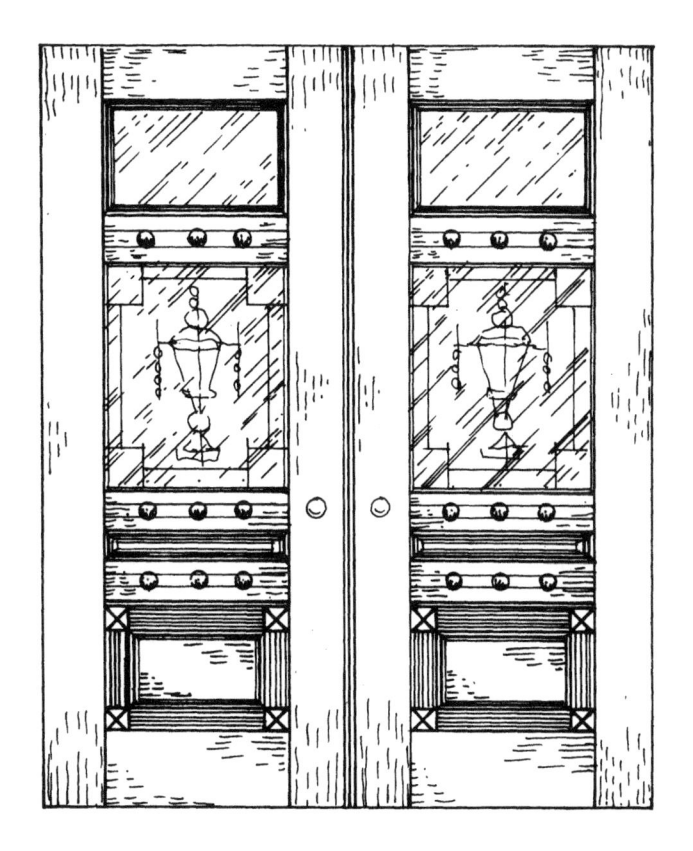

Door Designs • 91

DESIGN 21

These four doors came from an old catalog dated 1880. They were designed for a small brick and frame cottage and were described as "middle hall doors." That is, you walk in the front door, down the length of the hall, and then through the middle hall door and into the back kitchen or scullery. Such doors let the light through, yet they are semiobscured—so the maid can keep an eye on the front door without actually being seen. They are all framed and paneled with leaded glass above and plain moldings. The glass is colored vivid red and blue.

Door type and location: *Framed and paneled sash doors; good for a Victorian or Edwardian house, or would also look good in a modern house; best if the doors serve a vestibule or a covered porch.*
Construction and joints: *Traditional framed and paneled doors that use a good range of mortise and tenon joints. The windows are set into rabbets and held with plain molding.*
Skill level: *Easy doors to build. You have to start with the glazing and then make the doors to fit.*

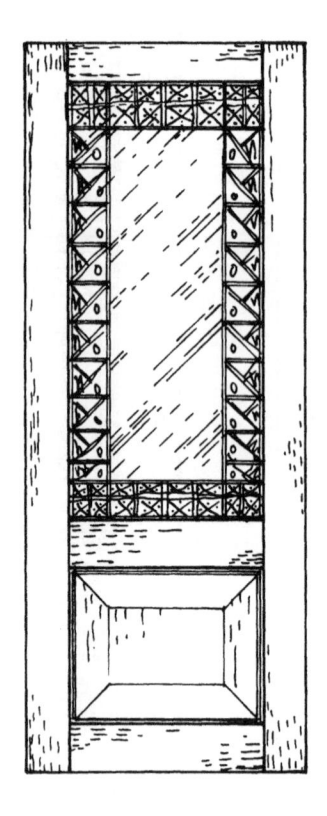

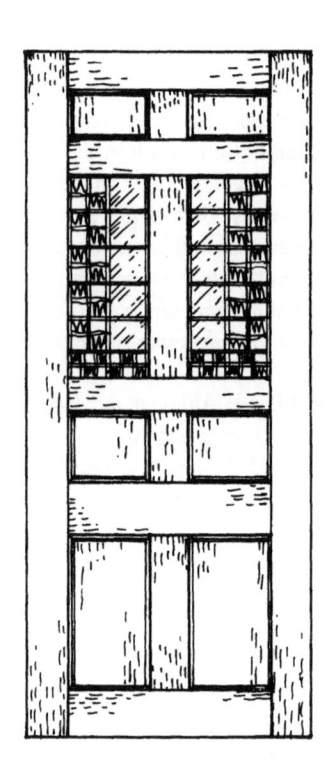

92 • Doors

DESIGN 22

If you have a passion for the Eastlake style and you are looking to build doors for a mansion, these are the doors for you. These doors were designed around 1880 for a California mansion. With their heavy machined embellishments and all the trim, they are designed to impress and stop you dead in your tracks. These doors make one think of bankers, railroad barons, opulence, and Henry James.

Door type and location: *Framed and paneled doors; good for a large Victorian house.*
Construction and joints: *Traditional framed and paneled construction, with the full range of mortise and tenon joints. There is a lot of fretting, mitering, and grooving, and the various embellishments need to be carefully housed and fitted.*
Skill level: *Although in many ways these seem to be difficult doors to build, they are pretty straightforward. The difficulty is not with any particular detail but rather with getting everything together to make a satisfactory whole.*

Door Designs • 93

DESIGN 23

Dignified, stylish, supremely elegant, and altogether practical, these vestibule doors were designed in 1880 for a brick and timber mansion to be built in Boston, Massachusetts. They are framed and paneled, with leaded stained glass above and heavy fielded panels below. Such doors would cost a fortune to build, but if you have a house big enough to take these doors, cost is probably not a concern.

Door type and location: *Framed and paneled doors with built-up curved heads; good for a Victorian house.*

Construction and joints: *Traditional framed and paneled construction, with the full range of mortise and tenon joints, plus some hammerheaded key joints for the curved heads.*

Skill level: *Difficult doors because of the curved heads and the need to deal with a window maker. Discuss the project with the window designer, then build the frames and have the windows built to fit.*

DESIGN 24

Well built and practical, this type of door is found in hundreds of thousands of homes across the country. This design may be something of a comedown from the Victorian doors on the previous pages, but if you need a plain and simple bedroom door—where the enclosed hall is short of light—this door is a pretty good solution. If you think it's too plain, build it in a fancy wood such as cherry.

Door type and location: *Framed and paneled door with a transom window above; good for an interior space where additional light is needed. The transom provides light without taking away from privacy.*

Construction and joints: *Traditional framed and paneled construction, with mortises and tenons.*

Skill level: *Easy to build—no more than a basic frame.*

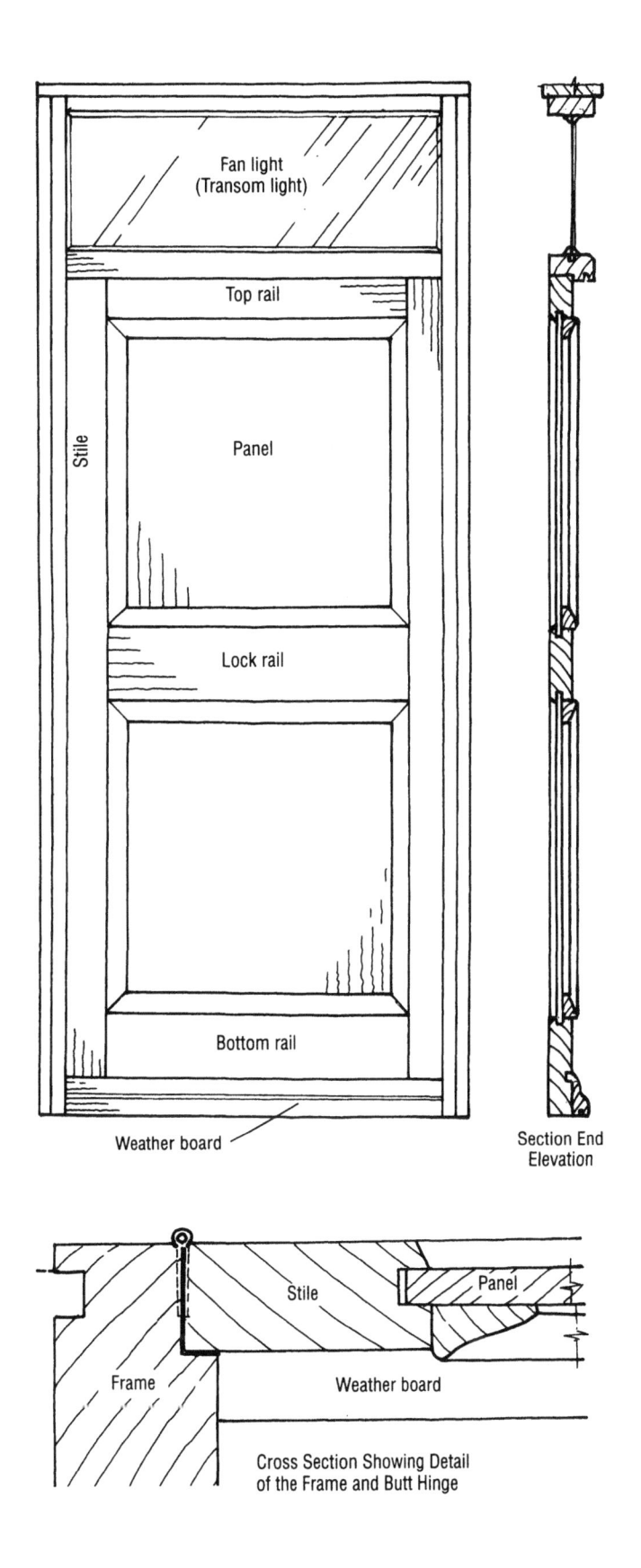

Door Designs • 95

DESIGN 25

Designed in the 1930s and described as "Jacobean," this door is built in oak. The moldings are worked from solid wood, and the whole thing is dark stained. It is well built, stylish, and modest, and it can be worked with a router. If your house was built between 1920 and 1930 and you want to rip out the horrible flush doors that misguided owners have put up over the years, try this door.

Door type and location: Traditional framed and paneled door, designed for a small between-the-wars house; could be used inside or under a covered porch.
Construction and joints: Traditional framed and paneled construction, with the panels built into the frame. You have to build in the panels when you are framing up the door.
Skill level: Tricky because of the need to get all the measurements just so. You'll need lots of trial put-togethers to make sure that the panels fit well.

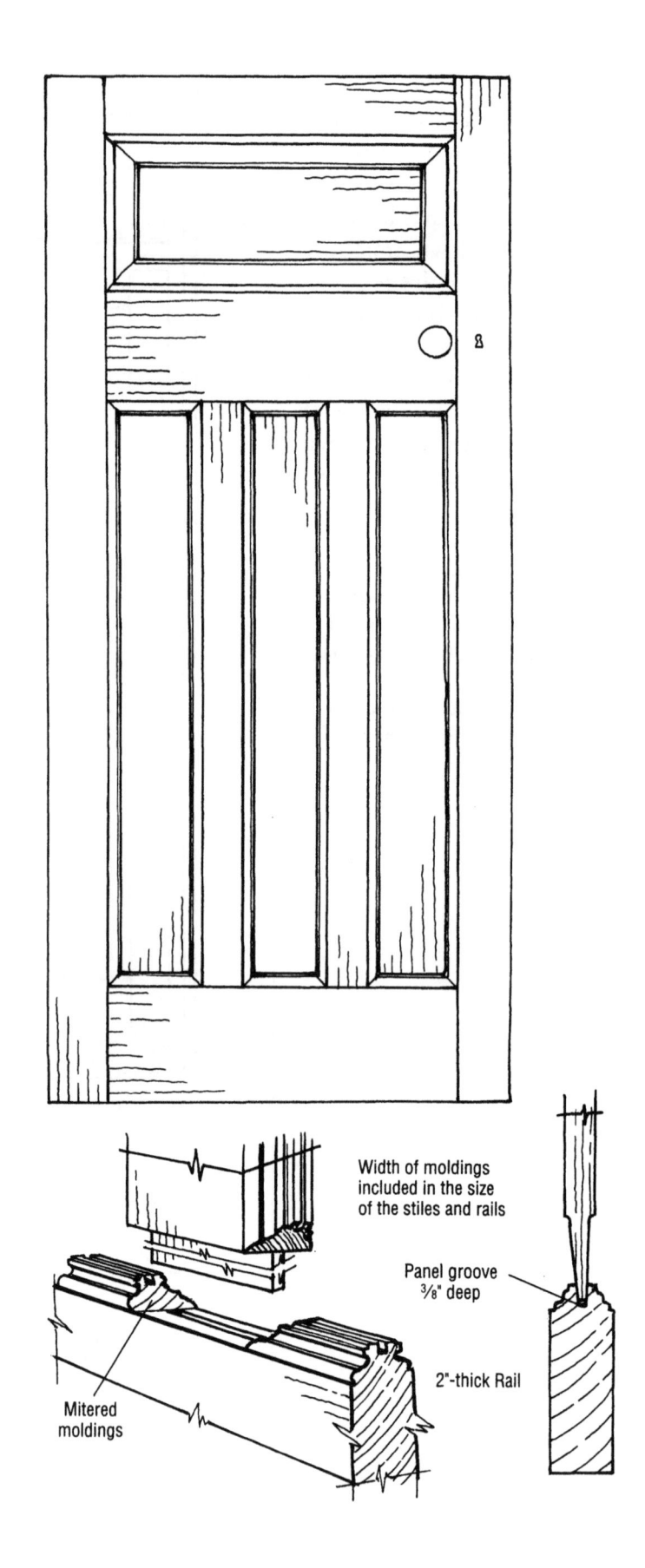

Width of moldings included in the size of the stiles and rails

Panel groove ⅜" deep

2"-thick Rail

Mitered moldings

DESIGN 26

Although this little 1920s front door isn't exactly pretty, it does capture the mood of the period. It would look perfect in the context of a small town house. The framing is worked with a groove and bead so that the finished door appears to have bead-flush paneling. The diamond decoration on the center panel is applied, and the central panel is embellished with a heavy bolection molding.

Door type and location: *A framed and paneled door; good for a small, tight, between-the-wars house.*
Construction and joints: *Framed and paneled, with mortise and tenon joints, plus stub tenons. The panels are built in at the framing stage.*
Skill level: *Although in many ways this is a straightforward door, the tightness of the design calls for a fair amount of precision. This door would be fun to make if you use a combination plane to cut the grooves and the beads.*

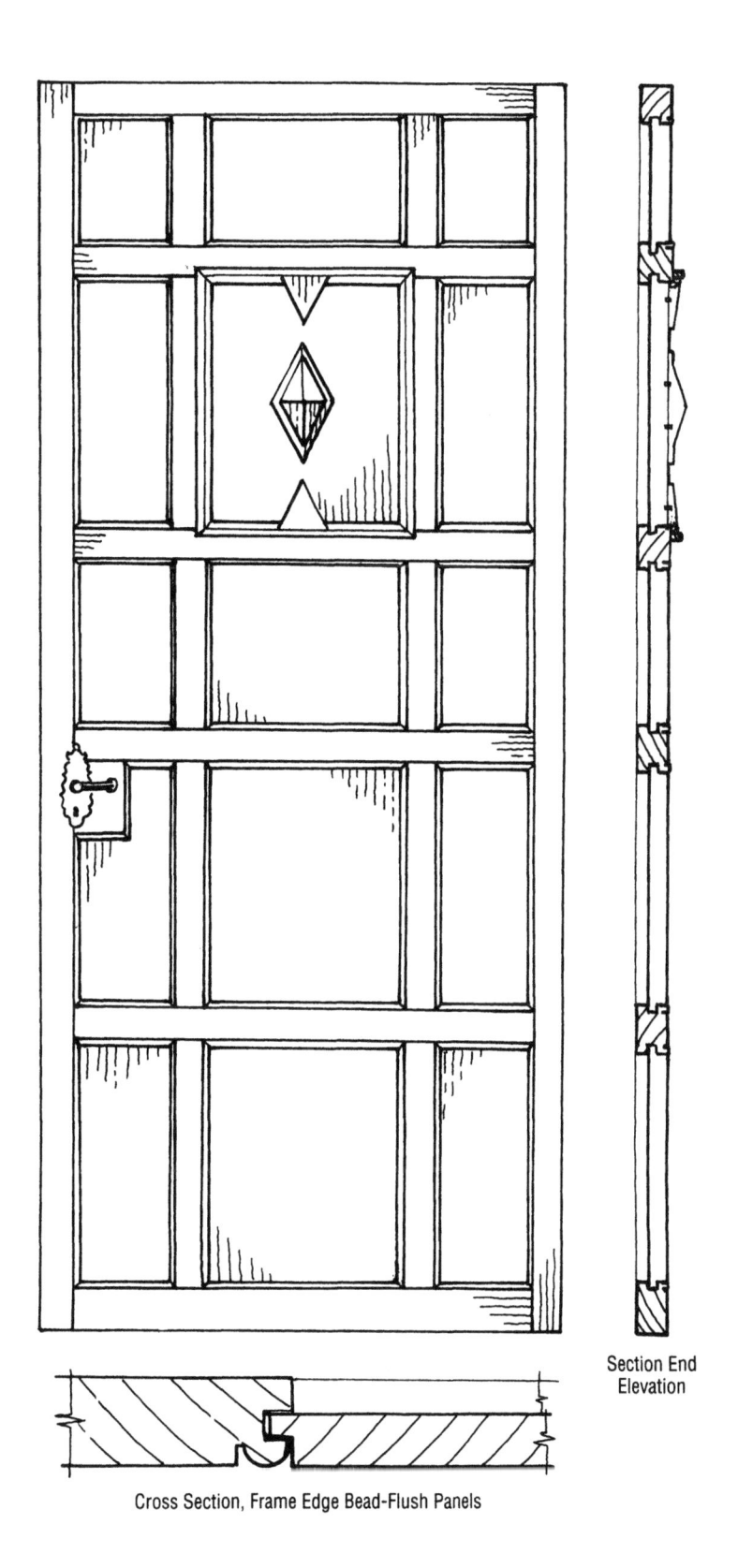

Cross Section, Frame Edge Bead-Flush Panels

Section End Elevation

Door Designs • 97

DESIGN 27

Described in an old catalog as neo-Georgian, these two doors were designed in the 1930s for a pair of small semidetached London houses.

Top: Framed and paneled with an arched top rail, leaded lights above, vertical muntins, and studded nail details.

Bottom: Framed and paneled with a glazed upper half; the glazing bars are in line with studded muntins.

Door type and location: Basic framed and paneled doors; good in 1930s and 1940s houses as a front door under a covered porch.
Construction and joints: Framed and paneled, with all the usual mortise and tenon joints. The vertical muntins are stub-tenoned into the rails. You could use plain moldings, or you could run grooves and build the panels in at the framing stage.
Skill level: Middling to easy, as long as you use an easy-to-work straight-grained wood such as American oak.

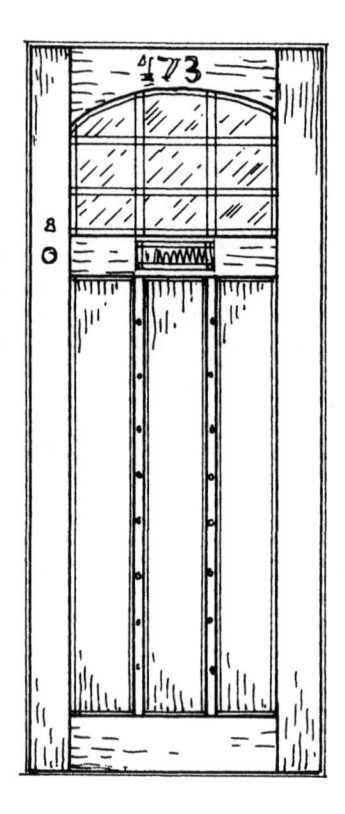

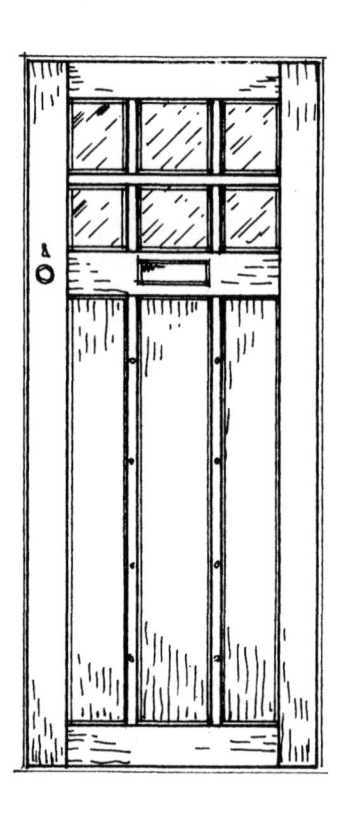

DESIGN 28

Although this charming Art Deco door is no more than a three-plank door, powerful imagery has been achieved simply by applying sawn sections to the central panel, cutting in the little windows, and running a drip bar around the top quarter. All in all, it's a beautifully stylish door.

Door type and location: Could be built as a three-plank door with modifications or with a solid flush-door-type construction; good in a period Art Deco house.
Construction and joints: Rabbeted butt joints, glued and nailed additions.
Skill level: An easy door to build.

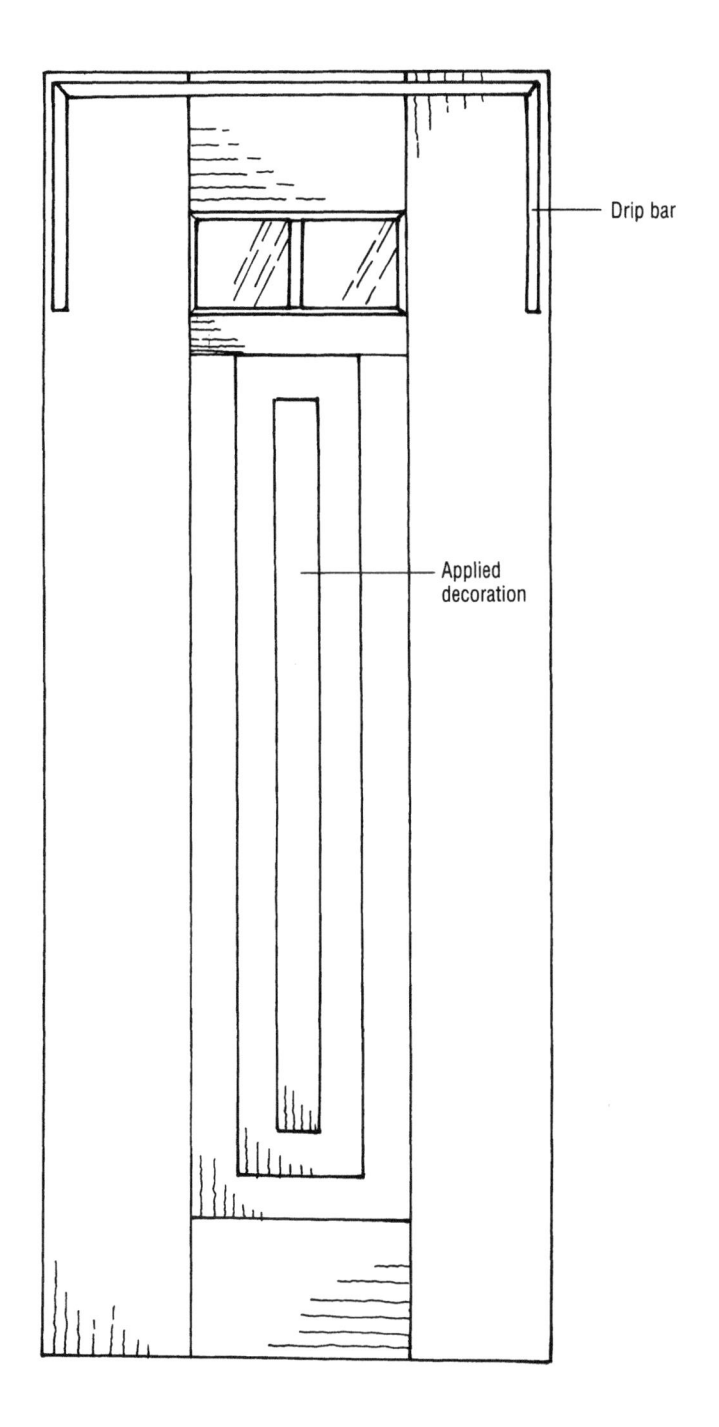

DESIGN 29

If you want to build a modern door, this group of doors, built between 1920 and 1950, may be the answer.

Top left: This neo-Jacobean door, built in 1930, is framed and paneled in oak, with a raised top panel and plain moldings throughout.

Top right: Built in the 1940s, this is a flush door with a pierced leaded light and applied molding strips to the face of the door.

Bottom left: An inner vestibule door dated about 1940, with slender glazing bars.

Bottom right: A rear entrance door built in the late 1950s, with horizontal glazing bars.

Top left:
Door type and location: *Framed and paneled; good in a between-the-wars house.*
Construction and joints: *Framed and paneled, with mortises and tenons and stub tenons.*
Skill level: *Tricky to build because of the need for precision.*

Top right:
Door type and location: *A flush door with applied moldings; good for a 1940s house, preferably protected by a porch.*
Construction and joints: *No joints—the whole thing is glued up as a cross-laminated slab.*
Skill level: *Easy.*

Bottom, left and right:
Door type and location: *Framed and glazed doors; designed as back doors serving the garden.*
Construction and joints: *Mortised and tenoned, with stub tenons for the glazing bars.*
Skill level: *Easy.*

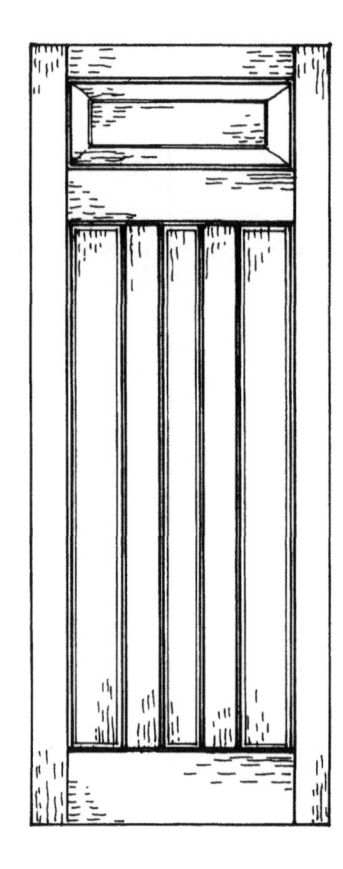

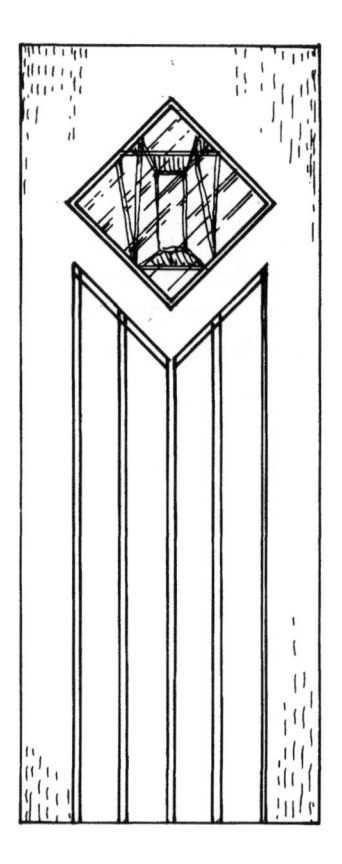

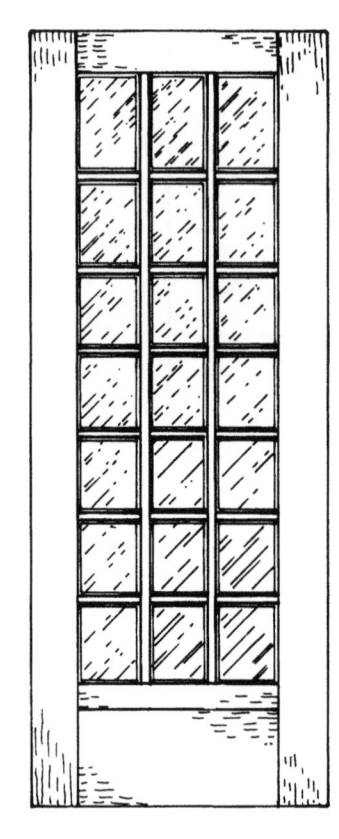

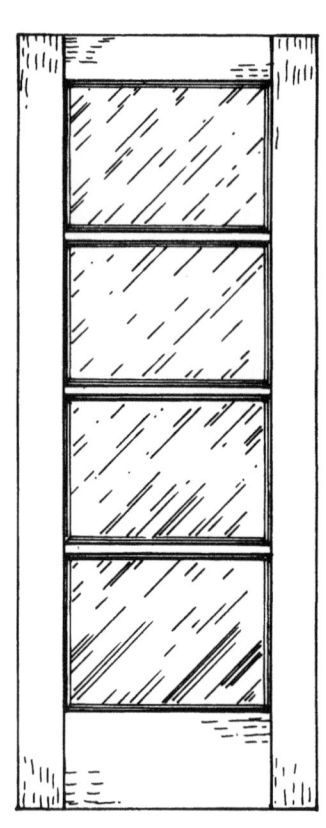

DESIGN 30

There is nothing romantic or wonderful about a pair of garage doors, but if you need a solid pair of doors to replace a tacky, modern, pressed steel up-and-over door or one of those horrible electrically operated roller doors, you can't go wrong with a traditional framed, ledged, and braced door.

Door type and location: *Framed and braced doors; designed for a garage.*
Construction and joints: *Framed and braced, using the full range of mortise and tenon joints—haunched and bare-faced.*
Skill level: *Relatively easy to build, but heavy to maneuver and hang.*

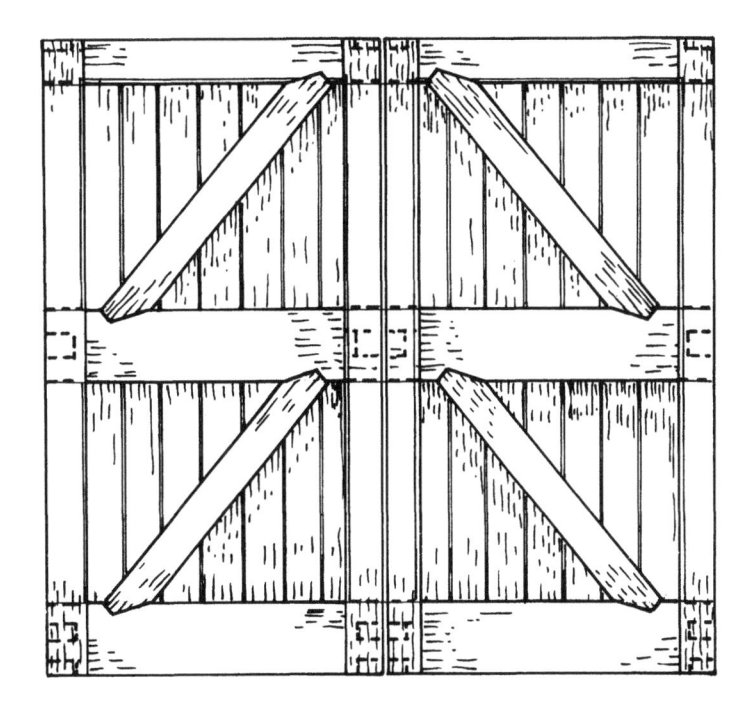

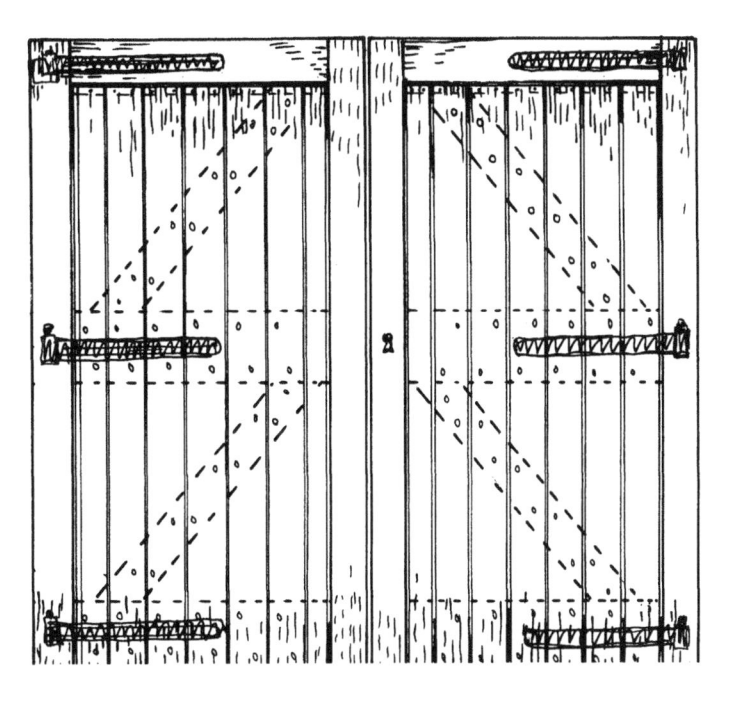

Part 2
GATES

Where would we be without all the fences, walls, trellises, palisades, and hedges that crisscross and parcel up our countryside? And what would all those boundaries be without the gates that give access to our neighbors? Maybe the country would be a friendlier place without high walls, barbed wire, and the like, but the reality is that without boundaries and barriers, we would all be in deep trouble. The plain and simple truth is that we all need our own little plot of personal space. We need fences to keep us in and them out, and we need gates for easy access. Whoever said, "A fence without a gate is a prison, while a fence with a gate is a paradise," certainly knew what he was talking about.

Over the centuries, our forefathers have come up with all manner of gates. There are picket gates, wicket gates, tollgates, farm gates, country park gates, cattle corral gates, wattle-hurdle gates, and railway crossing gates. There are gates with fancy top bars, tall gates that keep people from looking out and looking in, gates that shut themselves, and gates with spring catches. There are lych-gates, which are used during funerals as temporary shelters for the biers; there are kissing gates, bower gates that double as gazebos, and gates that allow easy passage for pedestrians but preclude horses and vehicles. If you want a gate—no matter the function or location—chances are that a gate has already been designed for just such a situation.

Although a gate should be pleasant to look at, it is most important that it be fitting and functional. A gate must open and close without undue hindrance, it must be strong and stable, it must be designed so that it stays put when fully opened or fully closed, and it must function without failing and without doing damage to life and limb. However, you also have to be mindful that a gate sends a message. It may say, "Please come in," or "Stay out!" or "Watch out for the dog!" or "Please close the gate behind you because there are horses in this field." The size, shape, and design of the gate say it all. For example, a low, white picket gate with an easy catch and flower borders to the side says, "Welcome." A tall, solid gate with strap hinges, a piece of razor wire on the top rail, and no visible catch most definitely says, "Keep out!" Yes, a gate needs to open and close, but more than that, the gate must send the correct message.

DESIGN AND CONSTRUCTION

Just as doors are traditionally named in accordance with their form and function, so it is with gates. For example, a gate might be described as a five-bar, an open picket gate, and so on. When you are considering making a gate, you must ask yourself the following questions:

- Do you want the gate to be welcoming, or do you want it to be intimidating?
- Do you want the gate to be a low, open, see-through structure, or do you want it to be a high, closed visual barrier?
- Does the gate have to keep kids or pets in or out?

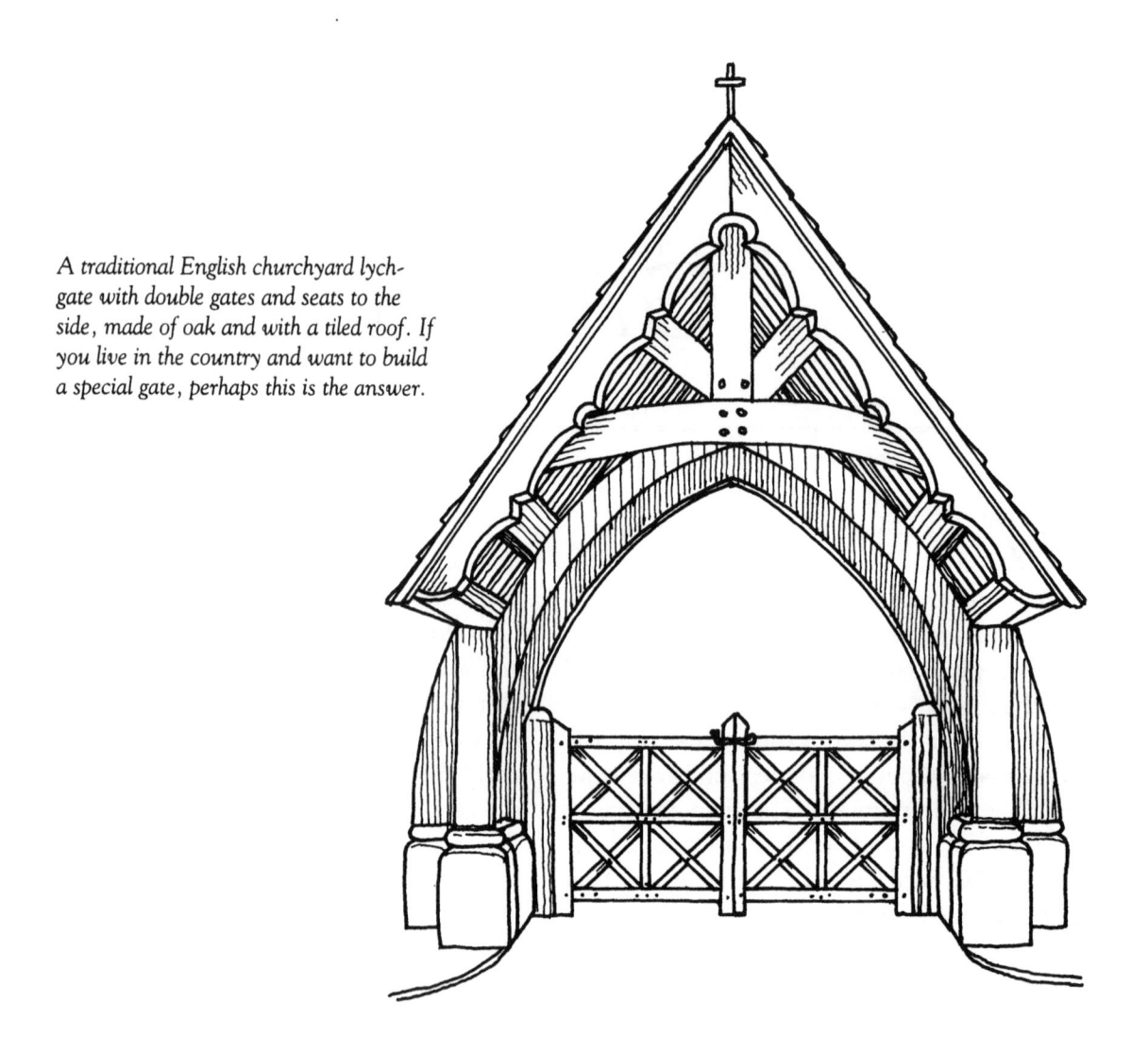

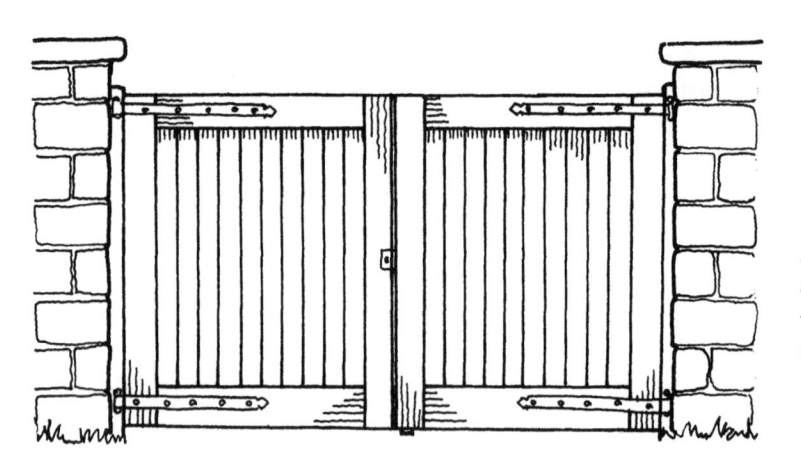

A traditional English churchyard lych-gate with double gates and seats to the side, made of oak and with a tiled roof. If you live in the country and want to build a special gate, perhaps this is the answer.

A courtyard gate provides both security and privacy. It's strong, it sends out the right message, and it's too high to easily look over.

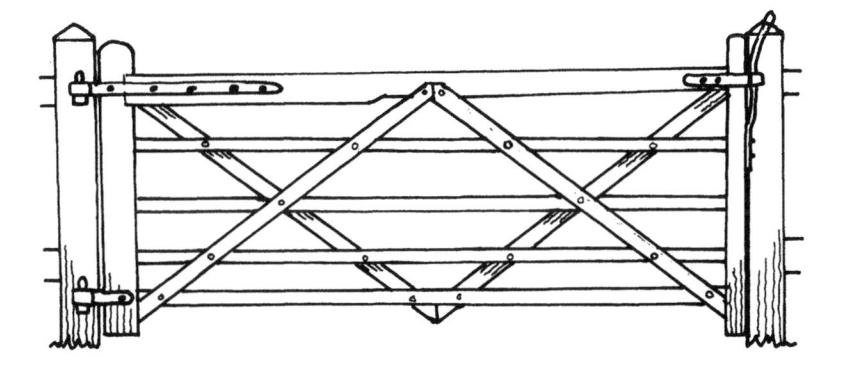

A traditional five-bar field gate is the perfect answer for keeping stock under control—horses, sheep, cattle, and the like. If you study the design, you will see that the gate draws its strength from the triangulation of the hinge stile, the top rail, and the brace. Note how the top hinge takes most of the weight.

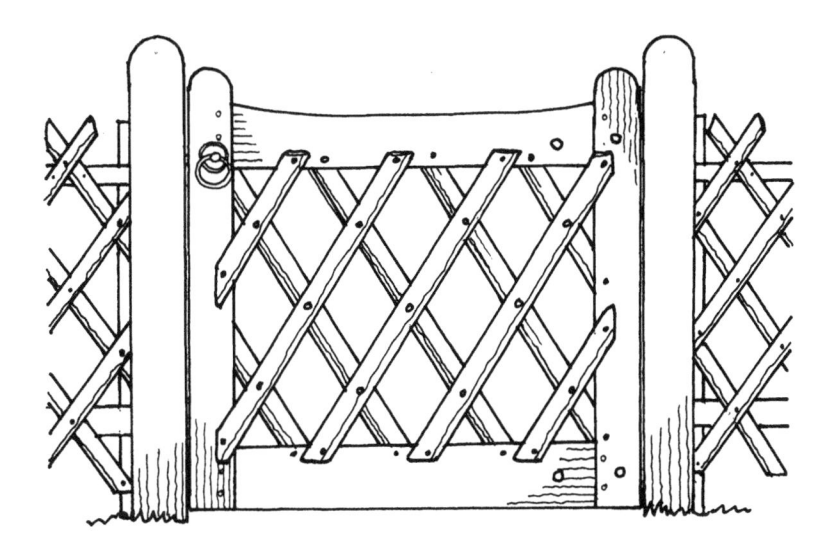

Though a small gate like this is designed to be inviting and friendly—perfect as the front gate to a country cottage—it's useless when it comes to keeping pets and kids under control.

- Does the gate have to keep stock in or out?
- Does the gate have to keep adult intruders out?
- Should the gate be decorative in the sense of being a roses-around-the-door type of structure?
- Must the gate be wide enough to take a vehicle—a car, a baby carriage, a bicycle, or whatever?
- Will it be a prestigious main entrance gate, or is it a modest side-access gate?
- Does the gate have to open in or out? This is especially important if you are setting a wide gate at the end of a short drive. If you have the gate opening inward, you have to be able to shut the gate once the car has been driven through.
- How will the gate affect your neighbors? Will it block their view? Will it send the wrong message? (A tall, intimidating gate might upset the neighbors.)
- Are there any local regulations that restrict the height of the gate?

Once you move a gateway, the flow of traffic will change accordingly. A new gateway can have a big impact on the way you and your family move across your property and the way other people perceive you and yours.

If you want to increase your privacy, route the kids away from your study, reroute cars, or make sure that your dogs stay put, a strategically placed

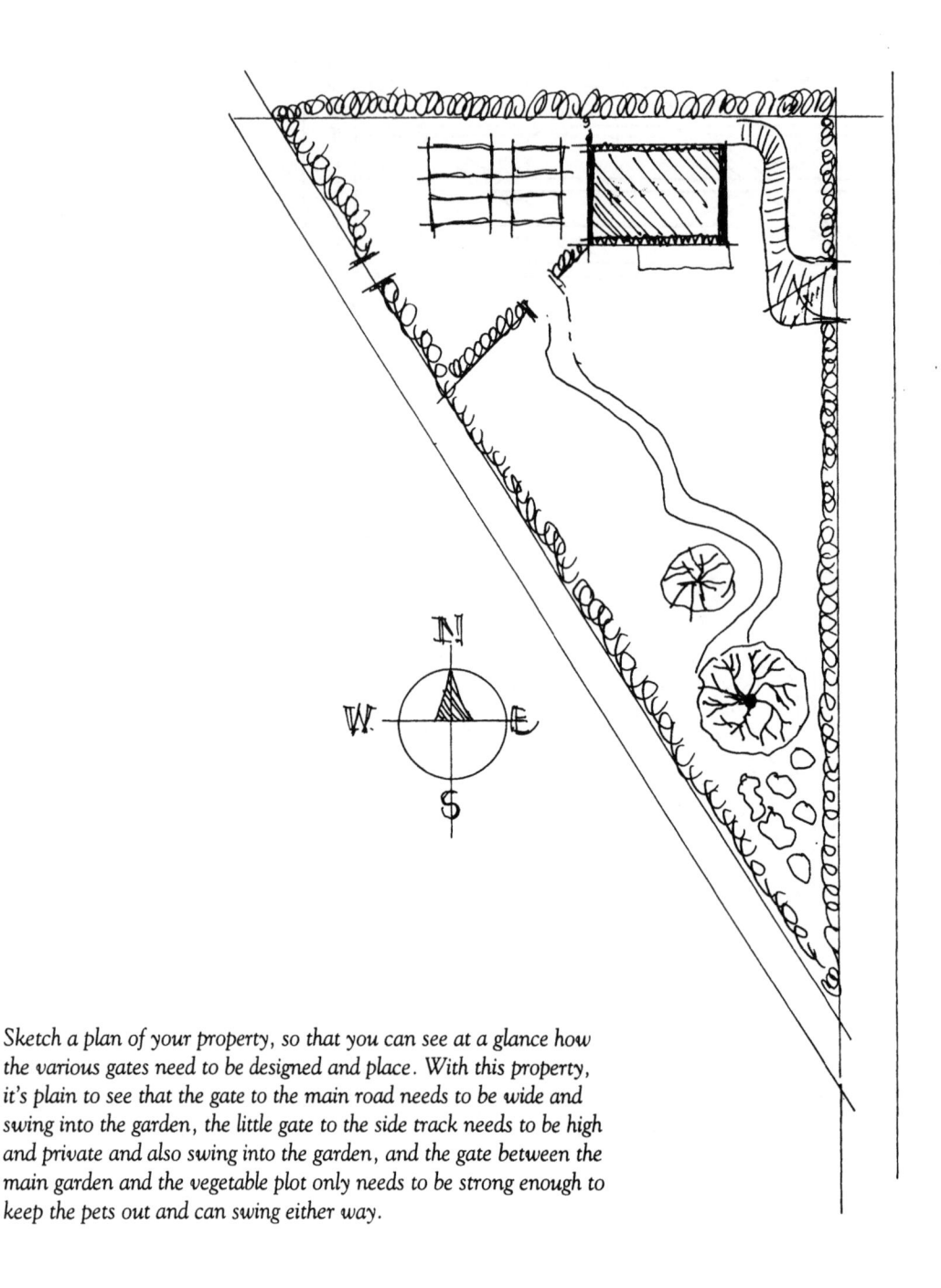

Sketch a plan of your property, so that you can see at a glance how the various gates need to be designed and place. With this property, it's plain to see that the gate to the main road needs to be wide and swing into the garden, the little gate to the side track needs to be high and private and also swing into the garden, and the gate between the main garden and the vegetable plot only needs to be strong enough to keep the pets out and can swing either way.

gate could be the answer. And one of the best ways to upgrade your home—with a view to selling—is to pay attention to the drive, the gates, and the front door.

The first step is to write up a list of your requirements. Then consider the size, design, and cost options, and draw a scale plan of your property. Set the whole thing out so that you can see at a glance what's going on in terms of traffic movement, views, and neighbors. By actually drawing in the position of the gate and its arc of swing, you will more clearly understand your needs.

Anatomy of the Gate and Gate Joints

If you're planning to build a gate, do you want it to be inviting and romantic? Or do you want it for privacy? Familiarize yourself with the parts of the gate—the joints and all the functional and decorative details—and you will have a clear understanding of your needs.

Frame

A basic gate frame has three components: the two vertical stiles, the two or more horizontal rails, and the one or more diagonal braces. Without the braces, the frame would rack and collapse.

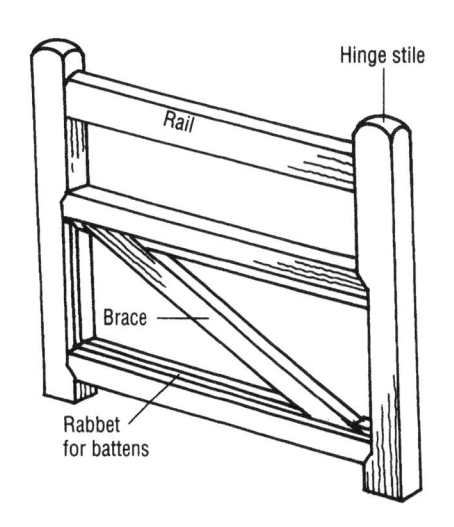

Picket

A slender vertical member used in the construction of some gates. Pickets tend to be 3 to 4 inches wide and between ½ and ¾ inch thick. The term comes from the Old French *piquer*, meaning to prick. Although the term has now come to mean a pointed stick or slat, it once meant the pointed part of a palisade or fence. To post a picket meant to put a guard at the gate. In England, the words *picket* and *wicket* are more or less interchangeable.

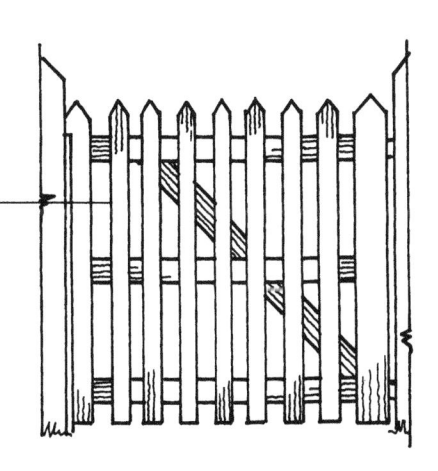

Head stile

The traditional name for the vertical member or stile that carries the latch. Other names are slamming stile, clapping stile, and latch stile.

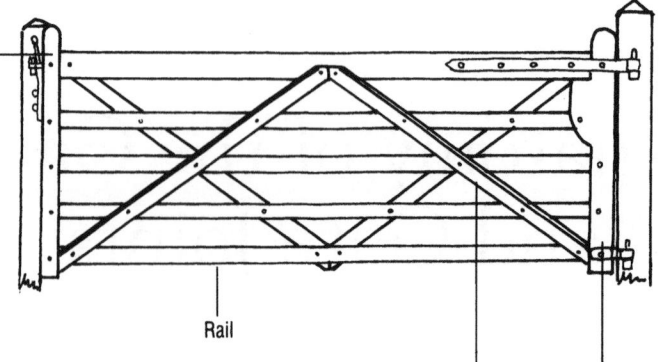

Rail

Brace

A member built into most gates, usually running from corner to corner to triangulate the frame. Unlike ledged and braced doors, which always have the bottom of the brace springing up from the hinge side, with gates (especially wide field gates), the brace springs up from the toe of the gate, with the top end being halved into the top end of the hinge stile. On this traditional English five-bar field gate, the braces are used to triangulate the gate and give it strength. The weight of the gate—that is, the weight of the top beam—is passed down through the main brace and on to the foot of the stile.

Harr stile (hanging stile)

Old English term for the stile that carries the hinges, meaning the stile nearest the hinge post. The harr carries the whole weight of the gate in its swing. A trussed harr is a stile that has an integral hewn bracket or ledge to support the top rail. This feature not only greatly enhances the appearance of the gate but also substantially increases its strength.

Curved stile

A design of hanging stile. The top curves over toward the gate, so that the top of the stile can carry the top end of the brace. The curved stile is the perfect combination of strength and decoration—it's strong, and it looks good. Traditionally, the gate maker searched out a curved piece of wood so that he could achieve the curve without cutting across the grain.

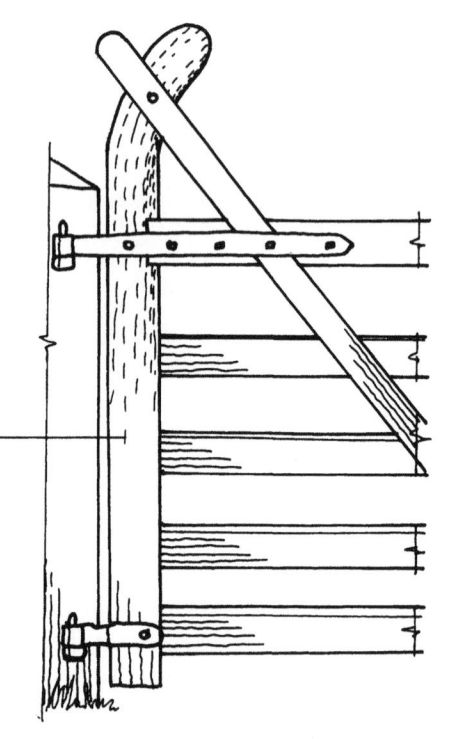

108 • *Gates*

Top rail

The topmost horizontal member; the primary member that triangulates by means of a brace through to the hinge stile. Traditionally, the top rail is much stronger than the other rails. Next to the hinge stile or harr, it is the most important part of the gate.

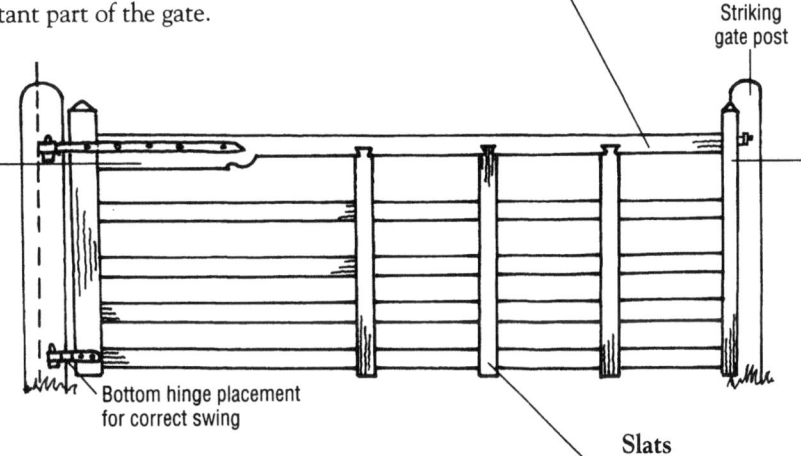

Striking gate post

Slam stile

Another name for the latch or head stile; also called the clapping stile. It is the stile that slams against the locking post.

Jowl

The decorative underside of the top rail on a traditional handmade farm gate. In Britain, it is possible to date a farm gate and make a good guess at where it was made by the design of the carved jowl. The jowl gives a taper to the top rail, so that the part nearest the hinge, which carries most of the strain, is heavier and more substantial.

Bottom hinge placement for correct swing

Slats

In the context of farm and field gates, the lesser vertical members between the stiles. In some parts of England, the slats are known as sloates or muntins. In small paneled gates, the small open vertical members are called mullions

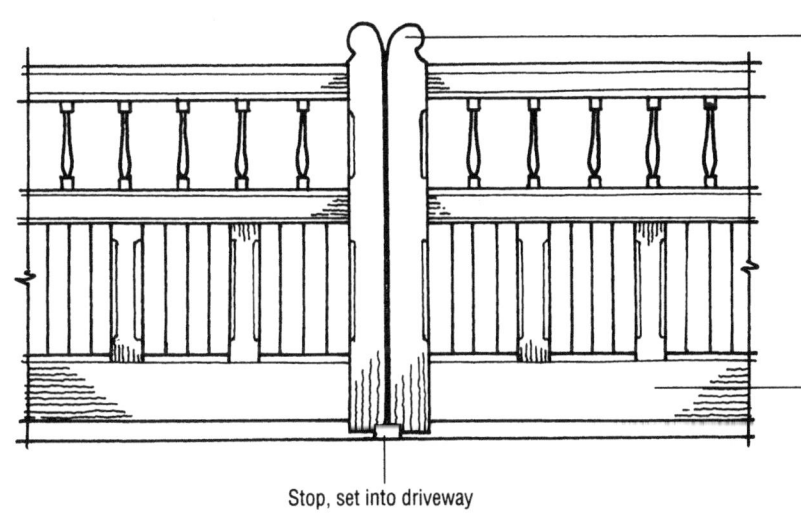

Meeting stiles

The two middle or mating stiles on a pair of carriage gates. One stile is held against a stop by means of a sliding tower bolt, so that, in effect, it becomes the gatepost against which the other stile is closed.

Stop, set into driveway

Baseboard

Sometimes called a gravel board, meaning the horizontal board that runs along the bottom of some gates, usually easily replaceable and designed to take the brunt of the wear.

Anatomy of the Gate and Gate Joints • 109

Stump latch

A small stump set low in the ground, used to hold the gate open. The gate is swung open so that the toe or some part of the bottom rail locates in the latch. To release the gate, you simply put your foot down on the hooked end. The good thing about such an arrangement is that it can be operated without bending down.

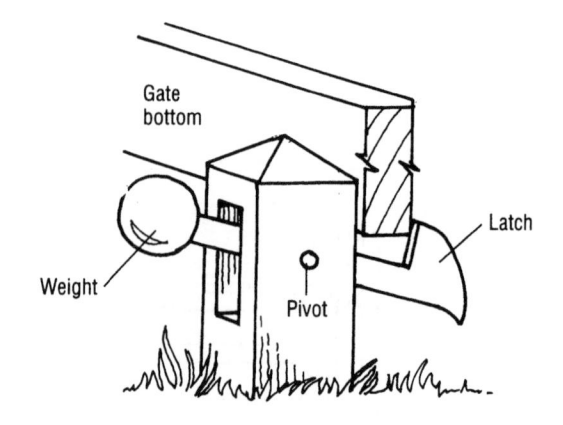

Hunting latch

Another name for the long sprung steel latch used on some farm gates. The latch gets its name from the fact that it can be easily operated from horseback. The hunting latch comes as a set of four components: a sprung steel arm, a bolt, a staple, and a catch. The bolt is used to fix the latch in position on the slam stile, the staple is screwed or bolted into the stile so that it contains and supports the latch—so that the latch doesn't get pushed out of place—and the catch is driven into the gatepost.

Vertical spring latch

In the context of field gates, a sprung steel bar that is bolted to the locking stile and locates in a staple on the locking gatepost.

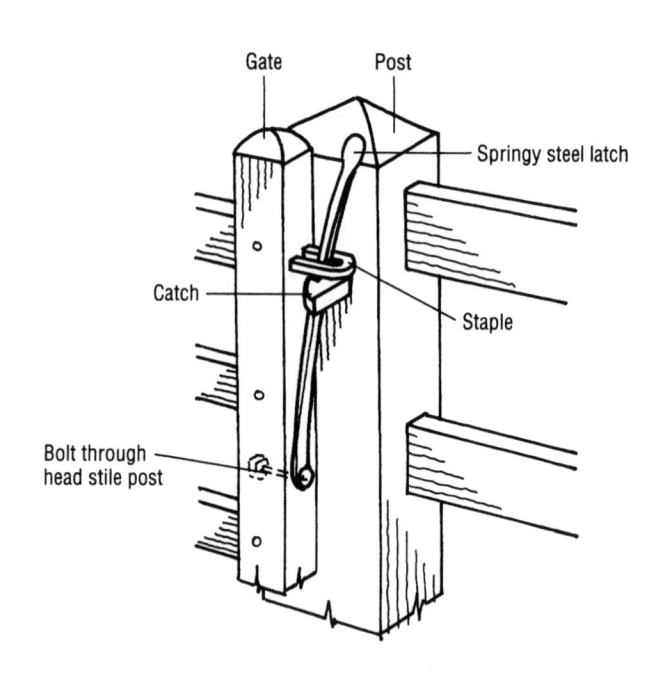

Hook and thimble

In England, the hook and thimble are the parts that make up the forged iron hinge of a traditional farm gate. The hook is the part that you drive or bolt into the gatepost, and the thimble is the part that is bolted and clinched to the hinge or harr stile. With a traditional farm gate, the bottom hook and thimble are either longer than the top ones or set to one side of the top ones, so that when the gate is hung it will remain in the fully closed or fully opened position. The idea is that if you let go of the gate, it will automatically swing open or closed, depending on where it is in the swing.

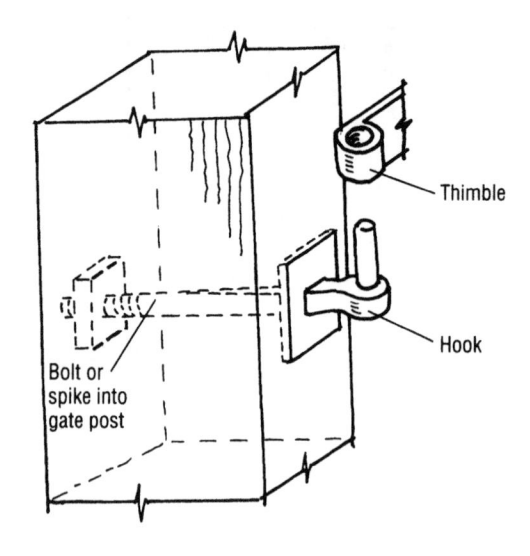

Ball finial

A ball capping used to decorate the top of the gatepost. Three classic examples of a ball finial are shown.

Left: A simple turned ball with a collar, mounted on a square section post.

Middle: A turned double ball and cap, mounted on a square section post.

Right: An integral ball that has been turned from the post.

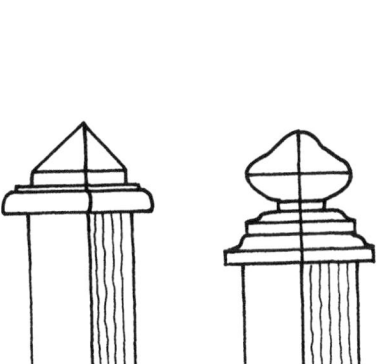

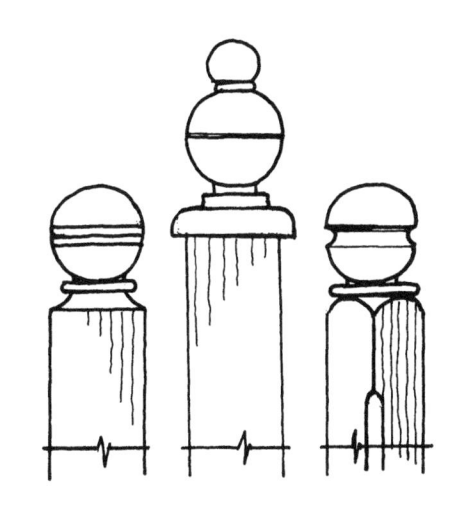

Post caps

Decorative caps used to top gateposts. They might be flat or shaped like balls, pineapples or acorns.

Pediment

A rooflike detail used to decorate the top rail of some gates.

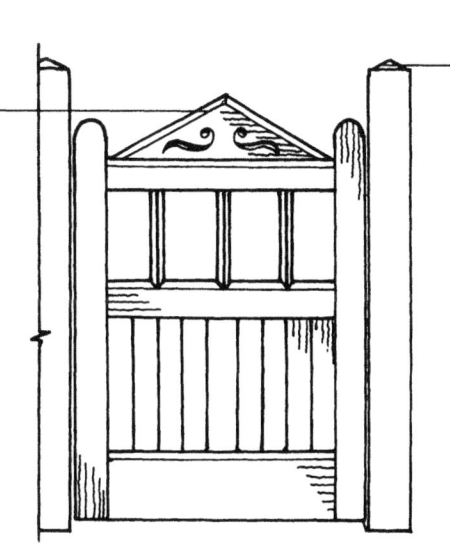

Pyramidal top

A gatepost design consisting of a square section post with the top chamfered back so that the top of the post looks like a pyramid.

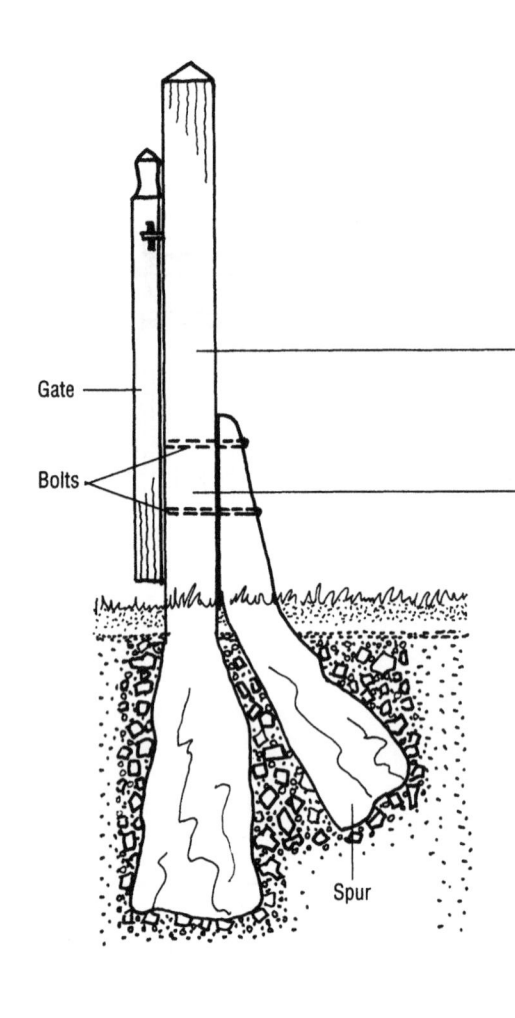

Striking gatepost

The post against which the locking or slamming stile of the gate comes to rest; another name for the clapping gatepost.

Spurred post

A spurred gatepost is one that has been strengthened by having an additional member abutted against its foot or butted and fixed with through bolts. With the whole works buried and tamped with earth and rubble rammed against the spur, the post becomes superstrong and stable. With a spurred gatepost, the spurred part needs to be buried within the gateway, to minimize the effect of the leverage when the gate is swung into the closed position. In the case of the shutting or locking post—also called a slamming or striking post—the spur needs to be buried on the opposite side of the post to the slamming side.

Butt

In the context of field and farm gates, the part of the gatepost that is belowground. With traditional farm gates, the butt is left in its natural, unsawed, straight-from-the-tree state. The design of the butt—its size and weight—is such that the post is going to stay put. It was reckoned that a well-planted oak post with a good butt could be left alone for the better part of a lifetime. A saying went, "Plant a gatepost with your grandfather, and mend it with your grandson."

Charred butts

Traditionally, the butts of gateposts were partially burnt, the idea being that the charred surface protected the bottom of the post from water rot and decay.

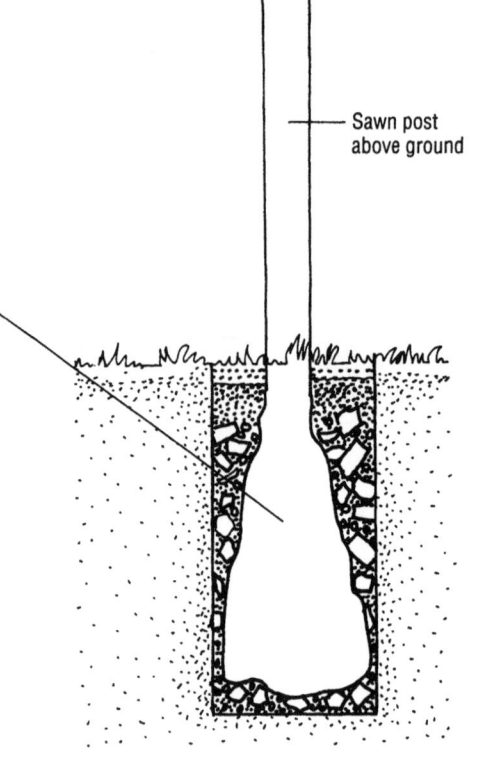

112 • *Gates*

Cross Section of Battens — Tongue

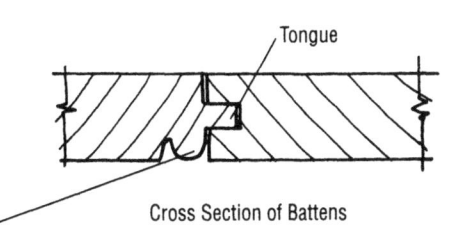

Bead

A small half-round molding used to decorate an edge, as on a gate faced with beaded tongue-and-groove boards. In this instance, the bead is used to decorate and disguise the tongue-and-groove joint. The bead draws your attention, so that you see the decoration rather than the joint.

Cleft wood

In the context of field and farm gates, the wood is split from the tree rather than sawed. It is achieved by driving a froe into the end grain and then levering it from side to side, so that the wood splits along the natural run of the grain. Other terms are *split* and *riven*. Gates that are made from sawed timber might look neater than their cleft cousins, but a gate made from cleft wood will outlast a sawed one by a couple of lifetimes.

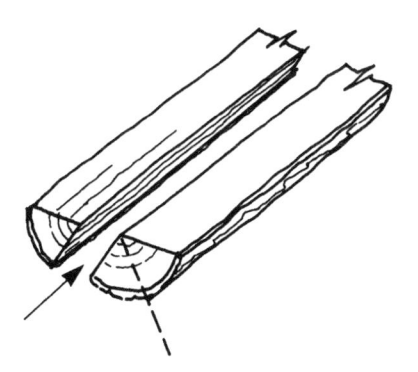

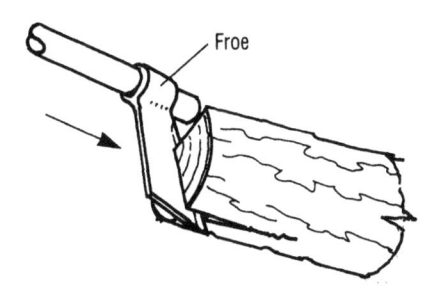

Footings

The rubble that is tamped in around the gatepost. Traditionally, the footings were a mix of broken brick and clay. Iron tampers are used to ram the footings in place. Field hands found that dry clay and rubble make the firmest mixture, because the clay swelled when it was damp.

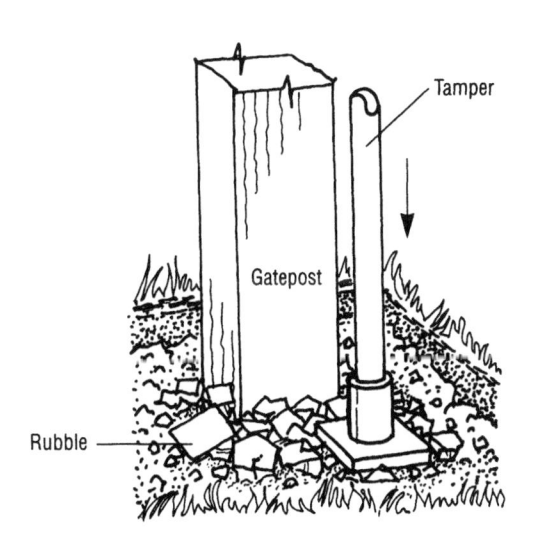

Anatomy of the Gate and Gate Joints • 113

GATE-MAKING JOINTS

Groove

A narrow slot or channel cut in the direction of the grain, usually around the inside face of the frame. Used with fancy gates to take the edges of panels or the ends or edges of boards.

Haunched and pegged tenon

The joint used on the harr end of the top of a five-bar field gate when extra strength and stability are needed. A mortise is cut through the harr, and then the top face of the mortise is enlarged to take the entire end of the top rail. When everything is in place, the joint is fixed with a wooden peg.

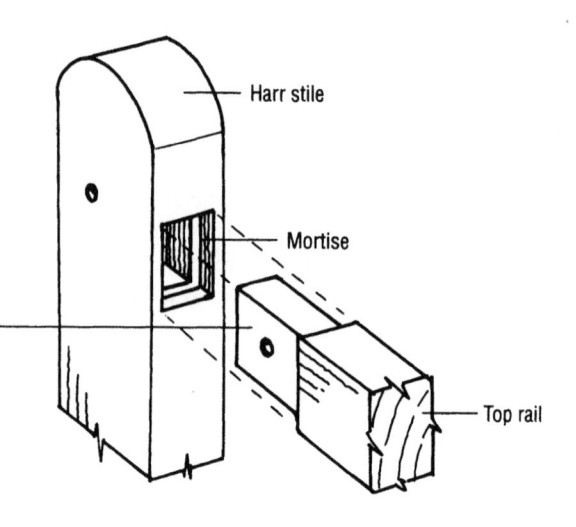

Lap joint

A joint used when the vertical slats or muntins are to be cut in flush with the top rail. Depending on the dimensions of the wood, the slats are halved and the top rail is notched.

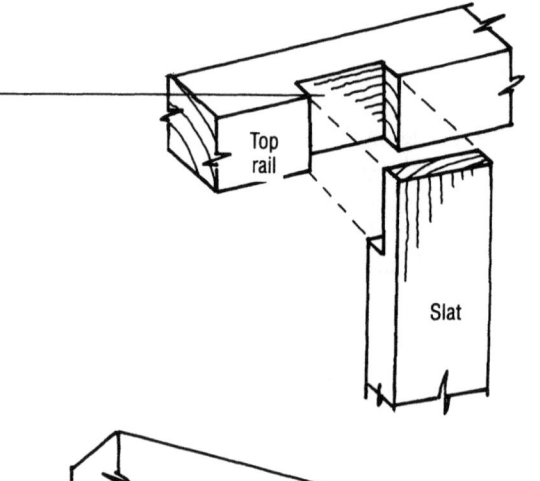

Muntin and brace tenon

A compound joint formed at the point where the slats or muntins meet the braces to be tenoned into a mortise that has been cut into the underside of the top rail.

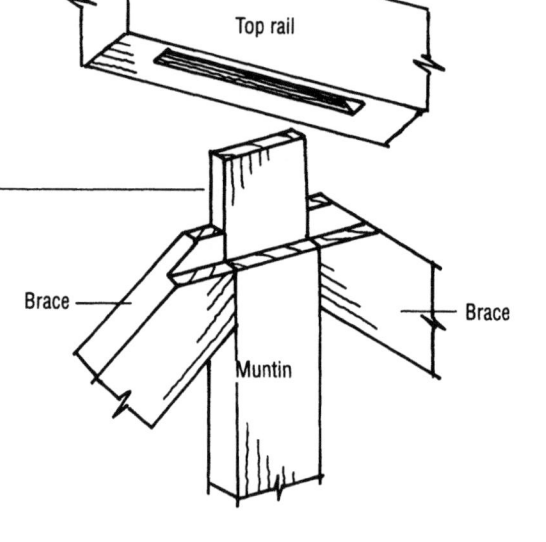

- Harr stile
- Through mortise
- Square peg

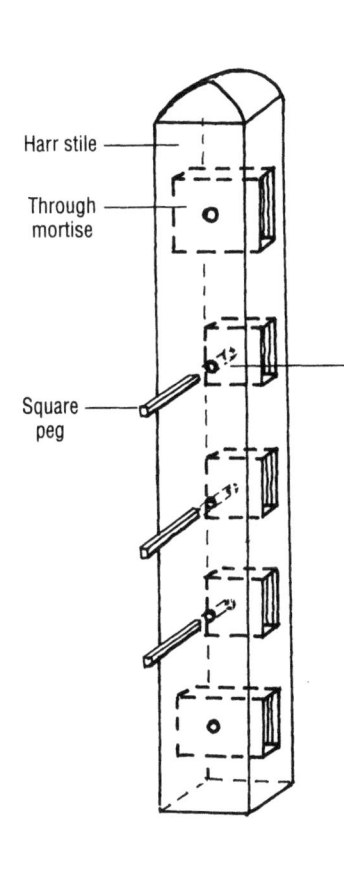

Pegged blind mortise

A mortise that stops short of the back end of the stile, with the whole joint fixed with a square section peg. Traditionally, with five-bar field gates, the three middle rails on the hinge side are blind-mortised into the harr post.

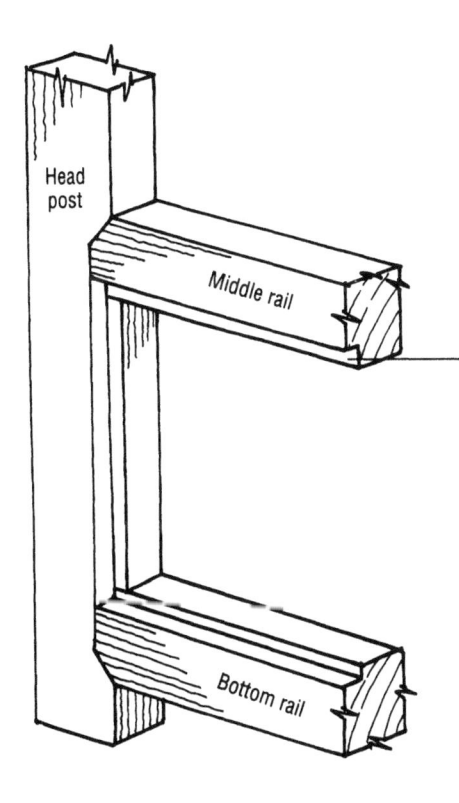

- Head post
- Middle rail
- Bottom rail

Rabbet

A step or recess cut on the edge of the wood, such as the rabbets cut on the edges of the stiles and the rails, into which the infill timber is nailed.

Anatomy of the Gate and Gate Joints • 115

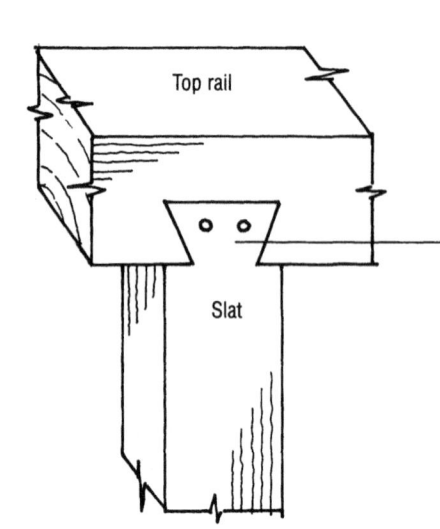

Slat dovetail

On a field gate, a feature in which the top end of the slats are dovetailed into the side of the top rail.

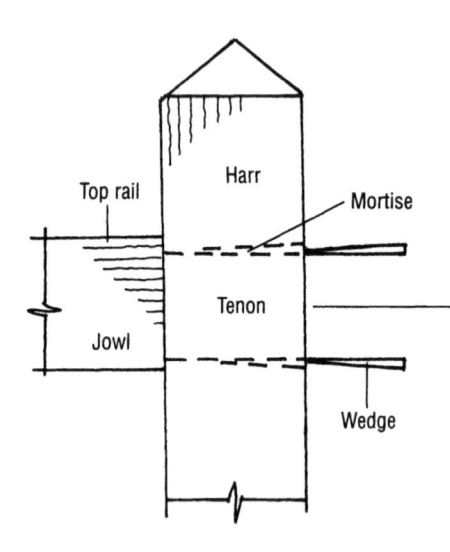

Tenoned and wedged jowl

A joint used in field gates when the large end of the top rail is jointed into the harr post.

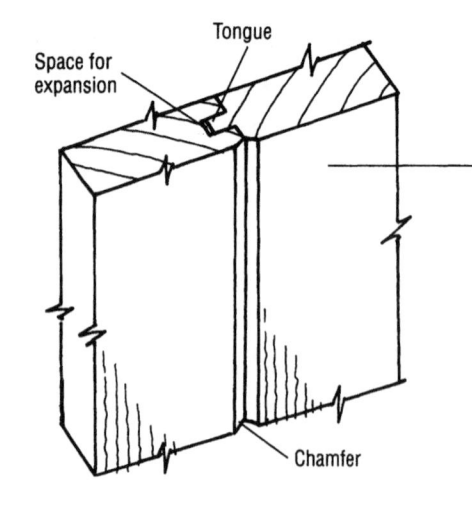

Tongued and grooved battens

A joint used in gates and doors for jointing the mating edges of boards and battens. In essence, one edge is cut with a tongue and the other with a groove. When they are slotted together, you get a good spread of solid surface to make a panel. And the tongue-and-groove jointing allows the whole structure to expand and contract without splitting or warping.

Generic Gate Types

BOARDED GATE

This simple board or batten gate has three ledges and a single brace—much like a ledged and braced door. With the boards being tongued and grooved and the gate being about 6 feet high, this gate is appropriate for a private backyard. The top of the gate can be cut straight, pointed, concave, arched, castellated, or whatever shape takes your fancy.

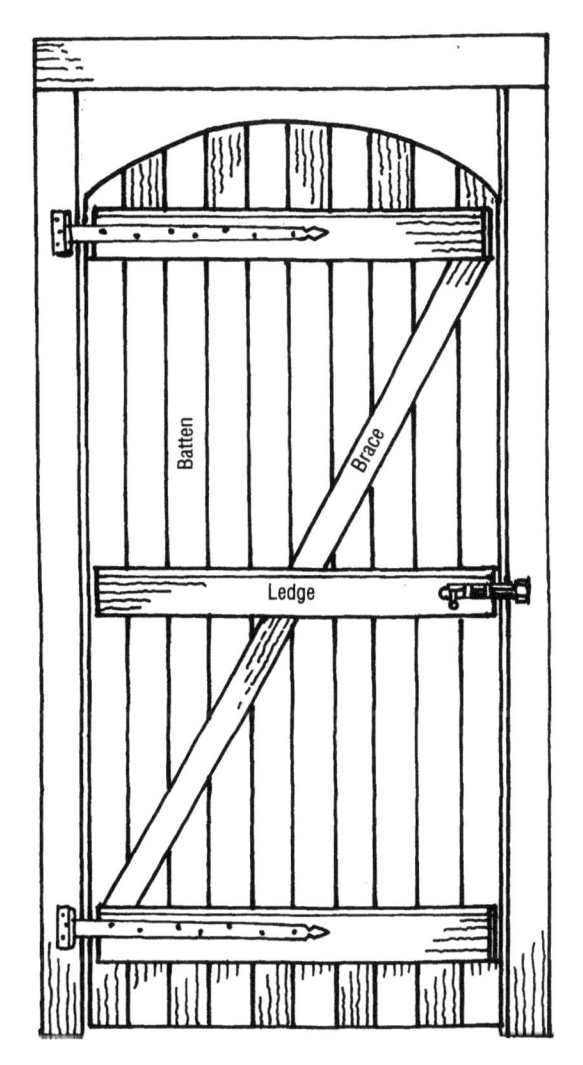

A close boarded gate showing its back face, with the brace running straight through from corner to corner.

BOWER GATE

A bower gate is a small, low gate that offers a shady retreat and is made from woven hazel rods—a bit like a woven sheep hurdle. Such a gate is romantic and rustic.

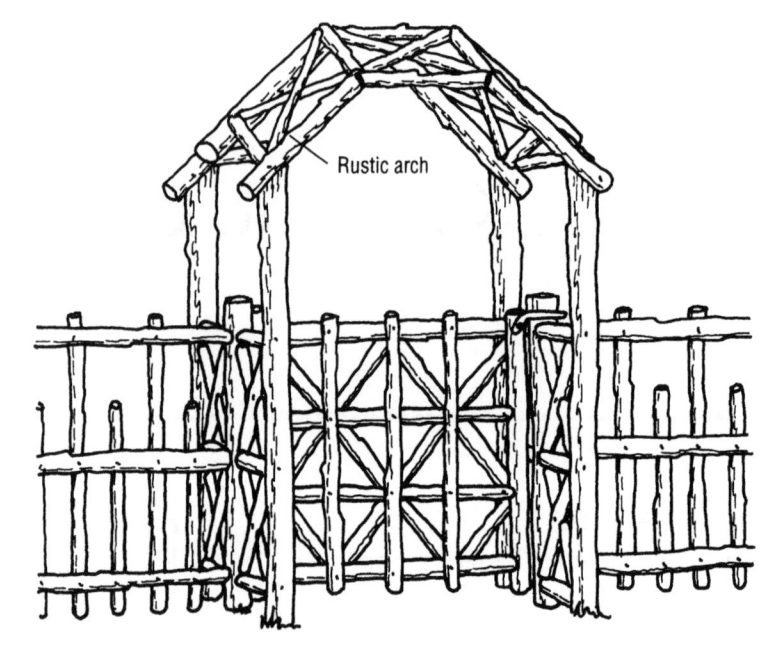

This bower gate is a real beauty. In this instance, the rods are left in the bark, and the whole thing is spiked together with nails. The arch is designed as a support for climbing plants.

CARRIAGE GATE

This pair of Victorian gates has finials on the gateposts and curved tops to the stiles. Each gate has two stiles and three rails; the bottom half is boarded, and there are secondary members.

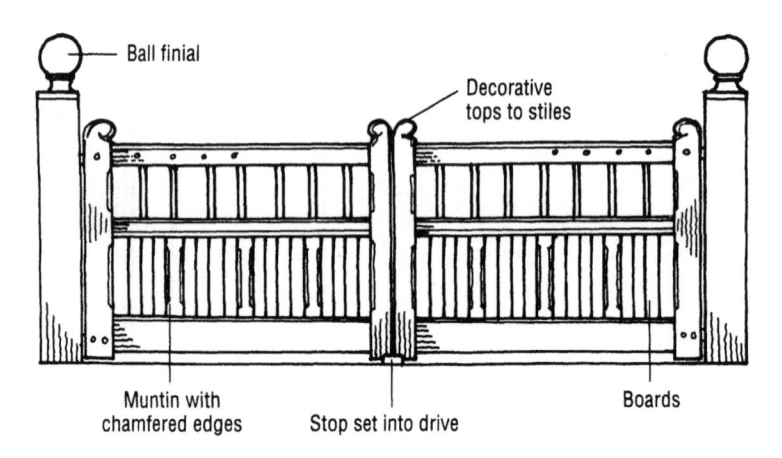

A carriage gate of this type is dignified and substantial. Note how the muntins are needed to brace the frame.

CARRIAGE GATE WITH COMBINED WICKET GATE

This wide-span gate is designed as a driveway gate with a small wicket gate to the side. The idea is that if you want to drive the car through, you open both gates; otherwise, you open only the wicket gate. The two gates total 9 feet wide, but they could be wider if you wanted to modify the design. These are traditional framed gates with stiles, four rails, and braces. The two bottom rails are covered with short palings to keep the kids and pets in, and the tops of the stiles are rounded and chamfered. If you need to cut costs, this is a good option, because you get two gates with only two gateposts rather than four.

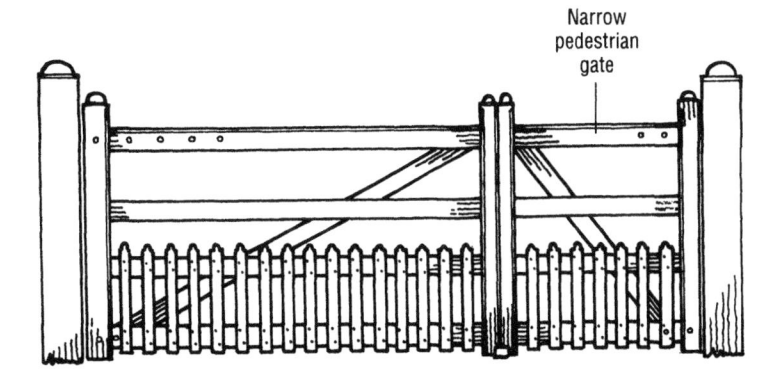

A carriage gate with a side wicket is a good solution if your driveway isn't wide enough to take two sets of gateposts. The only downside is the need for a tower bolt to hold the gate in place, and you have to bend down every time you open or close the gate.

COURTYARD GATE

The courtyard gate is actually a pair of gates infilled with vertical boards so that you can't see over. Such gates are usually about 6 feet high, with the shape of the top ranging from straight to convex or concave. If you have a small area that you want to keep private—perhaps a small courtyard in the city—this should be your first choice. If you want privacy plus security, you can top off the gate with a generous row of upended nail spiking.

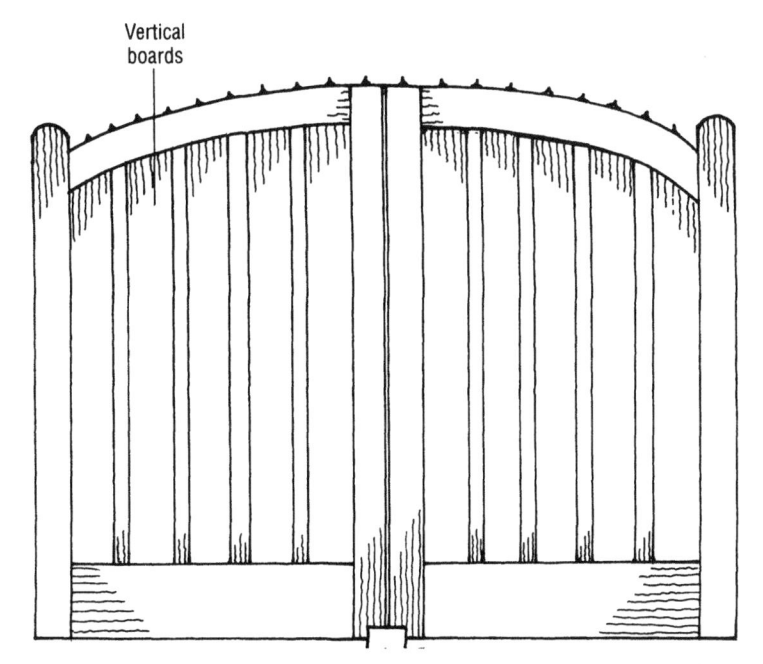

Courtyard gate, showing the front face. Note the extra-wide bottom rail or board and the row of spikes on top of the head. Besides giving the gate actual strength, the closeness of the muntins implies the strength of bars, portcullis gates, and the like.

Generic Gate Types • 119

CURVED STILE NARROW ENTRANCE GATE

On this small five-bar gate, the single brace drops down from the top of the curved hanging stile. This is a pretty little gate that draws its inspiration from large country park gates, but it would look good in a domestic setting.

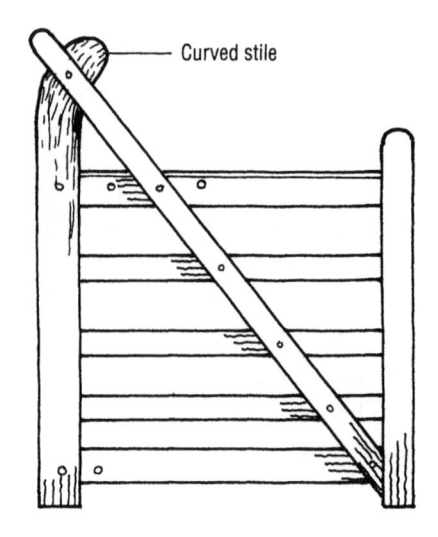

Maximum strength is achieved by having the curved stile cut from a found curved section, so that the grain runs around the curve.

EDWARDIAN BRACED AND BOARDED CARRIAGE GATE

This large carriage or drive gate has fancy posts and is framed up with two stiles, three rails, muntins, and braces; the bottom of the gate is boarded. Its main feature is that the braces are extended and shaped so that they look like Norman lancet arches—like church doors. Although the braces are decorative, they also create an immensely strong and stable form. If you are looking for a single large-width gate for your driveway and your house is Victorian or Edwardian, this gate would look fine.

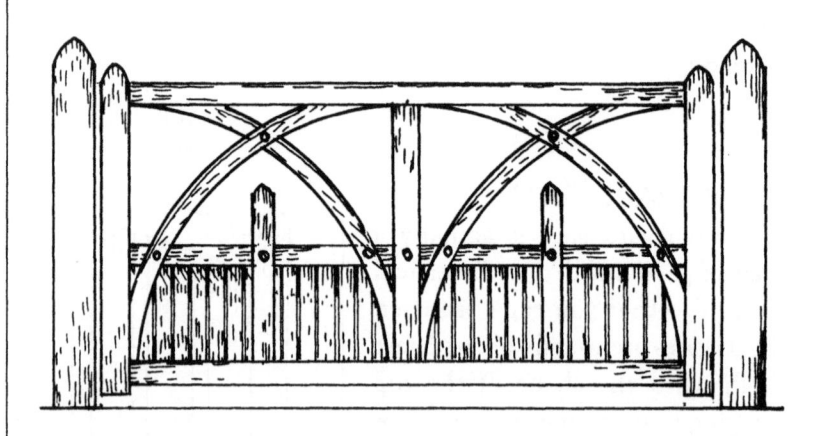

Braced and boarded carriage gate, showing the back or inside-garden face. With this particular design, the curved braces and the muntins and rails pass one in front of the other and are fixed with bolts. This maximizes strength by doing away with half lap joints.

ENGLISH COUNTRY PARK GATE

To make an English country park gate, take a traditional five-bar gate and add a brace that drops down from the top of the curved stile. There are many designs to choose from—some with vertical slats, others with cross braces, others with carved jowls, and so on. If you have a choice, make the gate in oak, and use the largest possible sections.

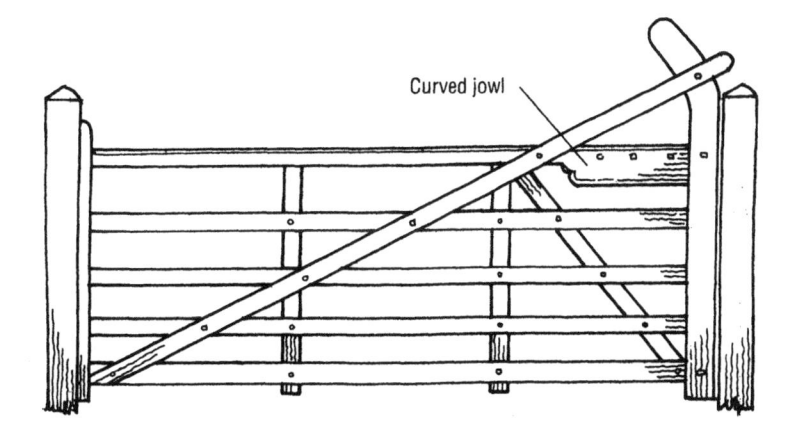

Note how the main brace, the curved stile, the top rail or beam, and the secondary brace all work together to carry the weight of this English country park gate. In this instance, there is no doubt that beauty is the direct result of form and function.

FIVE-BAR FARM GATE

Five-bar farm gates, sometimes called field gates, were traditionally used on English farms to provide easy access from one field to another. Most professional gate makers are of the opinion that the farm gate is the archetypal gate—the gate from which all other gates draw their inspiration. English field gates were traditionally made in three widths—8, 9, and 10 feet—but with today's larger tractors and combines, 12-foot gates and double gates are needed. For example, you might set two 10-foot gates together to create a 20-foot opening.

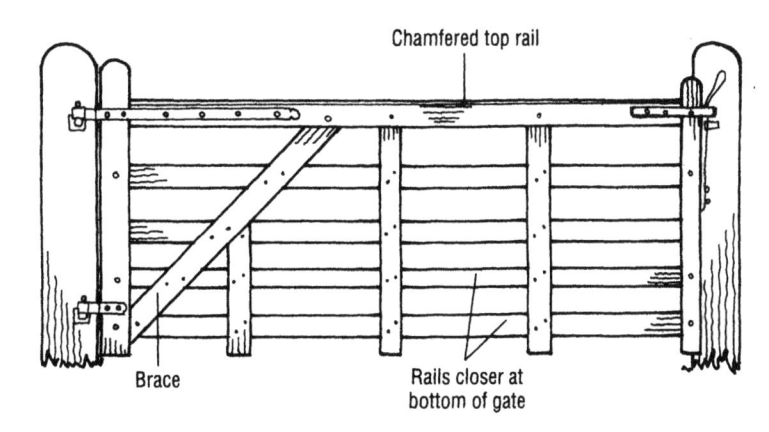

Farm gates were traditionally made from cleft oak and designed so that most of the weight was at the harr or hinge post end. Although there are many designs, they all have more or less the same basic form. The actual gate consists of two vertical uprights—the harr or hinge stile, and the head or slam stile. There are usually five horizontal rails—a massive top rail that tapers down from a section 5 by 5

A modern, factory-made five-bar farm gate. Note how everything has been reduced to the minimum—a straight top rail without a jowl or a taper, and all the rails set to the same spacing.

Generic Gate Types • 121

inches square at the harr end, plus four minor rails below this about 3 inches wide and 1 inch thick. The rails are arranged so that they are set increasingly closer together toward the bottom of the gate. The idea is that farm dogs can either jump over the gate or squeeze through the middle, but sheep and lambs are kept back. Depending on the district, the design of the top rail might be anything from a straight line taper to a taper that incorporates an ornamental carved shoulder or jowl at the hinge end. There might be one or more secondary vertical members or slats, plus one or more diagonal braces. These braces range in pattern from a single brace that drops down from the top of the harr to two V braces and four-brace double-X forms. The gate is hung between two massive square section posts that are set with about one-third of their height and half their weight in the ground. Even if you don't have a farm, such gates can be used as car or carriage gates and as the inspiration for small garden gates.

HUNTING GATE

The hunting gate is a small version of the field gate. It is about 5 feet wide and is hung in pairs. All the rails are mortised and tenoned, with the X braces being mortised into the underside of the top bar, and all the intersections of rails and braces being bolted through.

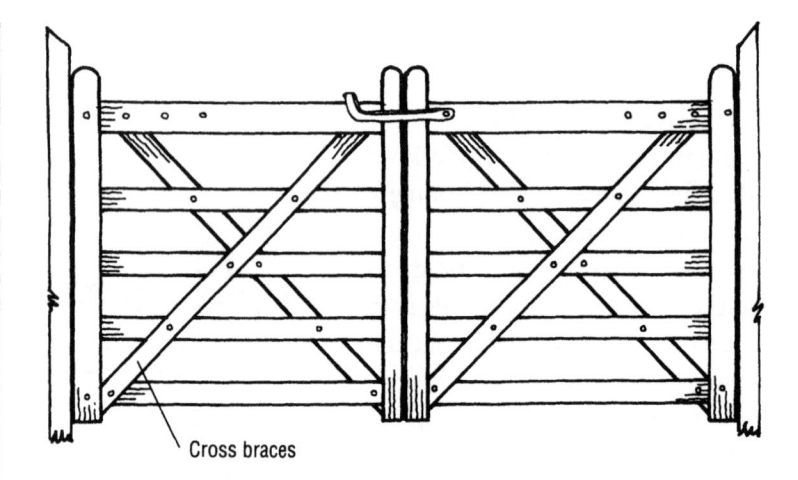

The whole idea of the hunting gate is that the hunter can stay mounted while opening the gate. There are no springs or complicated latches—all you do is lean down and slip the hoop-catch over.

KISSING GATE

The kissing gate is a small gate that swings freely within a small angled or semicircular enclosure—sometimes called a swing wicket or turnstile gate. The walker pushes the gate to the far side of the enclosure, stands in the space between the enclosure and the gate, and then moves the gate to one side and walks through. The idea is that the gate provides easy passage for walkers but denies access to bicycles, baby carriages, cattle, and the like. As to how the gate got its name, there are two schools of thought. One says that if two people try to use the gate at the same time, they get pressed together; the other says that there was once a tradition that allowed couples to kiss at such gates.

LYCH-GATE

From the old Anglo-Saxon *lic*, meaning a body or corpse, a lych-gate is a churchyard gate with a roof, under which the coffin is placed while the introductory part of the service is read. All that said, lych gates are now commonly built over domestic gates—the idea being that while they provide shelter for the mailman and for neighbors to chat, they also create a romantic image.

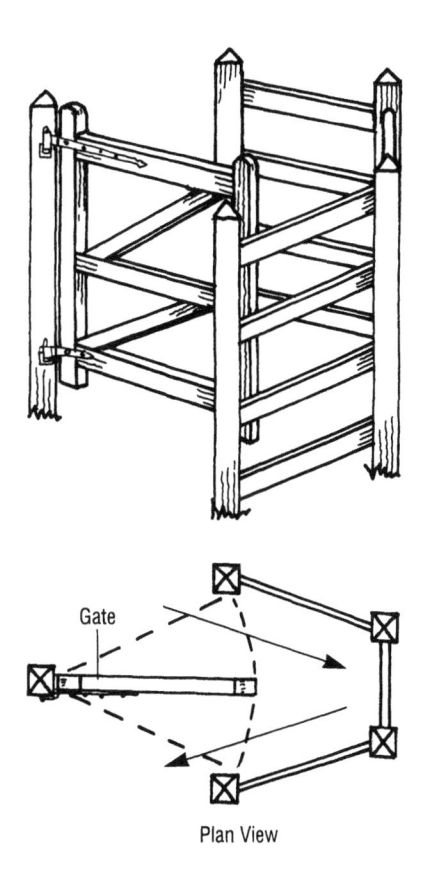

The kissing gate is perfect for a country footpath where you want to bar everything except pedestrians and dogs. It denies access to bicycles, horses, and the like. And if you keep livestock, you don't have to worry about the gate being closed.

A small domestic lych-gate. Note the classic design features—the jowls under the top rails, the symmetry, and the brackets under the roof.

Generic Gate Types • 123

NEW GARDEN ESTATE GATE

This high-framed garden gate has capped pillars to the sides, two stiles, and three rails, with the top rail curving down from the hinge side. The bottom of the gate is boarded, and the top is open and made from slender mullions. If you want a top-quality framed gate that will keep kids and pets in and strangers out—while allowing you to see through the top of the gate—this is a good option.

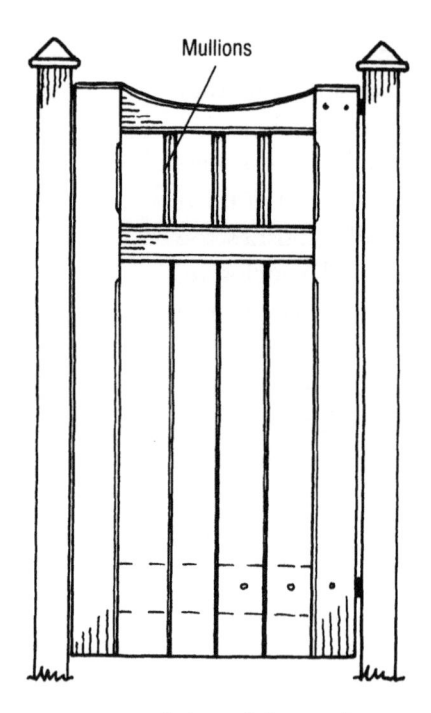

New garden estate gates can easily be scaled up or down to suit your needs. They can be tall, so that the mullions are at adult height, or they can be at a child's height.

PALE DIAMOND LATTICE GATE

This is a simple framed gate with stiles, three rails, and counter-diagonal pales on both sides of the frame. Although this type of gate can be made from sawed and planed sections, it's best if the frame is planed and the latticed pales are half-round in section and made from debarked forest wood.

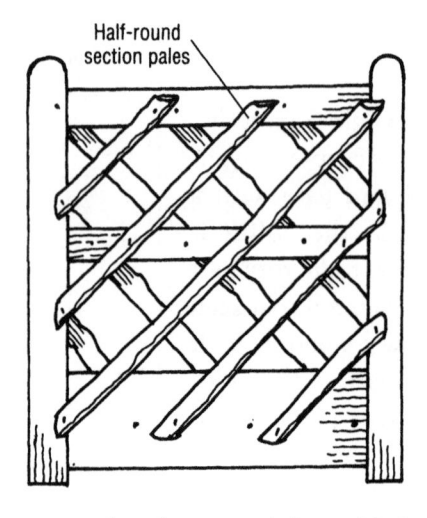

A classic country gate such as this can easily be modified to suit your environment. The pales can be cut from half-round sections, waste pieces with wanes, branches, or whatever happens to be at hand.

PALISADE PICKET GATE

The palisade picket gate is made from open-spaced pales or pickets, with two or more ledges and a brace. It is very much like a battened and ledged door, the only difference being that the boards are spaced rather than butted edge to edge. The tops of the pickets can be cut straight, round, or pointed or worked in any number of fancy shapes. This is a charming little gate, perfect for a country garden.

RIVEN GATE

Although just about any gate made from riven wood (wood that has been split rather than sawed) might be termed a riven gate, in England, it generally means a hurdle gate. It can be made of just about any wood—ash, willow, oak, or whatever happens to be available. A riven hurdle gate certainly doesn't provide the windbreak of a woven wattle gate, but it is stronger. The two side poles or stiles are mortised and pegged to take six or seven horizontal bars or rails. There is a vertical slat or muntin at the center and two diagonal bottom-to-middle braces, and the whole structure is fixed with clinched nails. Usually, the gate is made over a template or jig, with a small adze-like tool being used to cut the mortises. Although such gates were traditionally used for containing sheep, they are more commonly being used in domestic gardens.

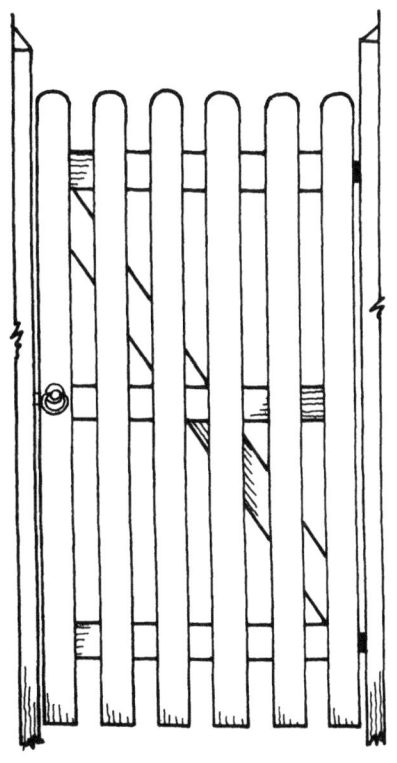

Palisade picket gate, showing the front face. The battens need to be about 3 inches wide and between $1/2$ and $3/4$ inch thick—the taller the gate, the thicker the battens.

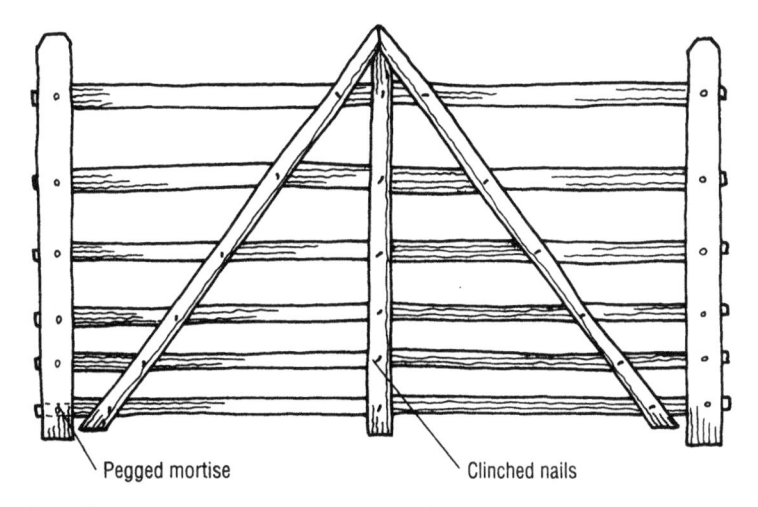

Although riven gates, sometimes called bar hurdles, were traditionally used for penning sheep, they now tend to be used in the garden to keep kids and pets in check. The strength and lightweight structure can be achieved only by using riven wood. You can't use sawed wood. It's important that the cleaving follow the natural lines of the wood, leaving the strength of the grain fibers intact. Sawing cuts across the fibers seriously weakens the structure.

Generic Gate Types • 125

RUSTIC GATE

This small cottage gate is made in the coppice poles tradition—meaning that the wood is still in the bark. The simple frame is made in much the same way as a hurdle gate. The two stiles are mortised to take four round-section rails, three vertical poles are nailed across the rails, and then the resultant spaces are set with counterchange diagonal braces. The end result is a quaint, rustic gate—one that brings to mind kitchen garden cottages and old folks with bonnets and clay pipes. If you live in the country and you like a simple, rustic feel, this is a good gate to make.

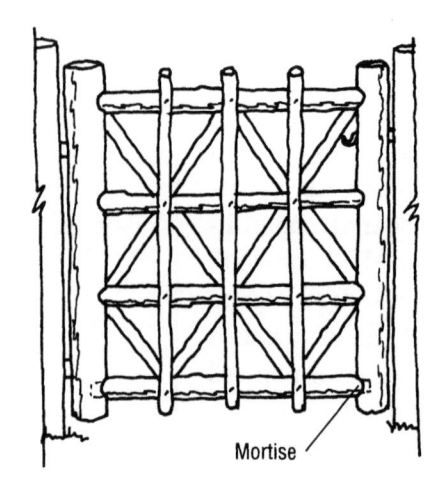

Although this rustic gate is easy to make, you have to be careful not to split the wood. It's best to drill pilot holes before nailing and then clinch the nails over on the other side.

SAWED FINIAL VICTORIAN COTTAGE GATE

This framed picket gate has band-sawed finials on top of the gateposts. The framing is made from two stiles, two vertical muntins, and three rails. There are pickets between the muntins and a small amount of fretted infill under the top rail. The top rail itself might be pierced.

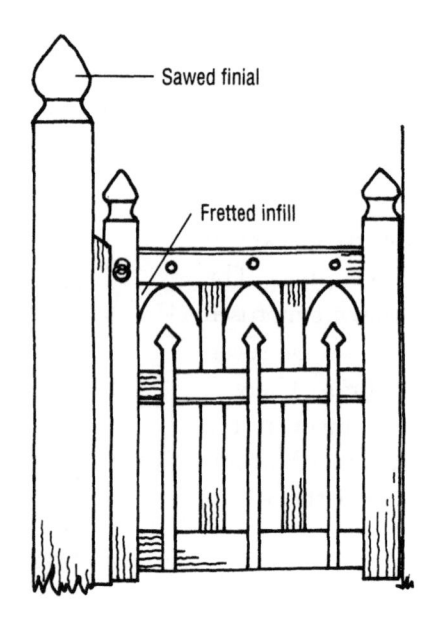

The fancy tops (finials) are achieved by using a band saw. All you do is cut the square section posts as seen in one side view, and then turn them over and repeat the same profile in the other side view.

SELF-CLOSING GATE

Such a gate might be anything from a gate with a metal spring to a gate that is hung wide at the foot so that it closes on its own. In the country, it is common to see field gates with lengths of chain running from the slam stile to the hinge post, with lumps of old iron hanging on the chains. The weight of the iron pulls down on the chain and keeps the gate closed. You could update this idea and work out something with weights and pulleys.

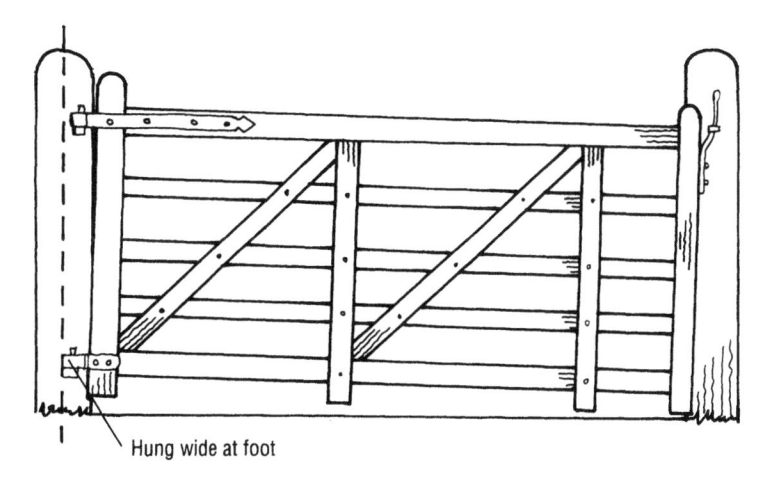

With this gate, the self-closing function is achieved by setting the top hook hinge slightly to the right of the centerline, so that it is nearer to the opening when the gate is closed. The toe of the gate rises as the gate is opened, so that if you let go of the gate, it will always slam back to the closed position.

SIDE OR SERVICE GATE

These gates are not main entrance gates but are reserved for domestic staff and tradespeople. This fact suggests that they are designed for the large houses of the wealthy. Such gates are usually well made and reserved in style. They are high, framed, double gates made in oak, with the frame paneled. The top rails are shaped a bit like a Tudor arch, to give the gate a grand feel. The panel is made from butted boards and grooved muntins, and the whole face of the gate is studded with rose-head nails.

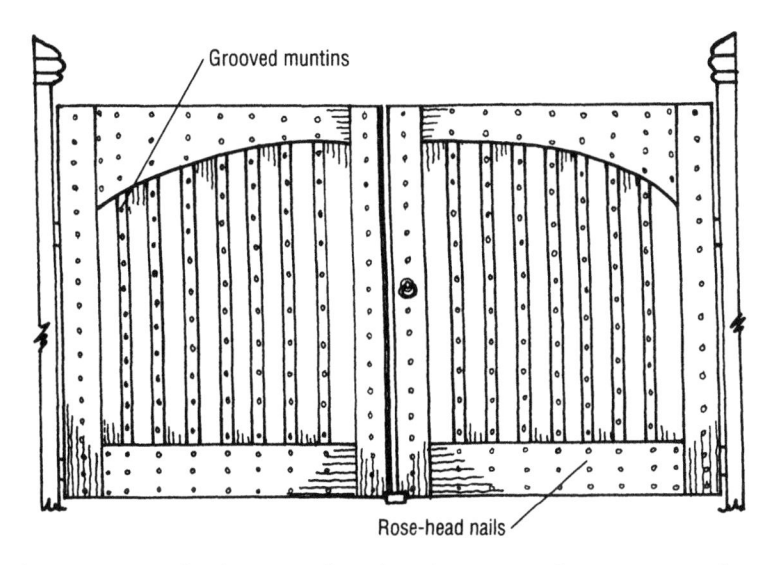

Service gates tend to be strong, formal, and sometimes plain aggressive—lots of dark oak and iron bolts and bars—to make them intimidating.

Generic Gate Types • 127

STILE GATE

The idea behind this rather curious mix of a stile or low fence and a small swinging gate is to keep out lambs and the like yet be low enough for a person to step over. You open the gate, step over the stile, and then shut the gate behind you.

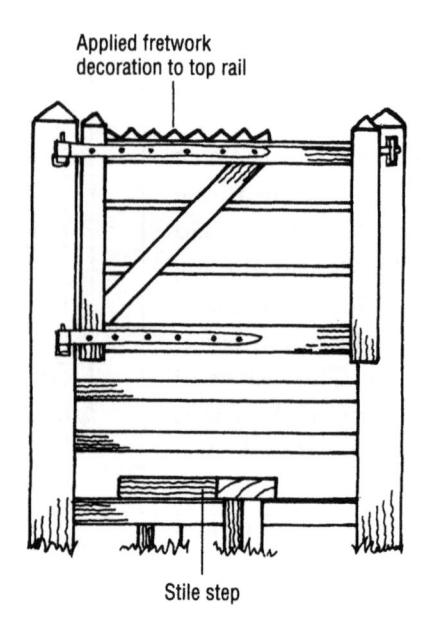

The stile gate is a belt and suspenders answer to livestock and the worry about the gate being left open. Even if the gate is left open, the bottom bars and the narrowness of the opening will deter most field animals.

SUSSEX FIELD GATE

Although a Sussex field gate is much the same as any English five-bar field gate, the details are more complex. The top bar or rail is tapered with a massive decorative jowl, there are three vertical members instead of one, and the diagonal brace rises at a sharp angle from the foot of the hinge or harr post to finish at the top of the middle vertical member. All rails and uprights are mortised and tenoned, and the underside of the top rail and the brace are chamfered.

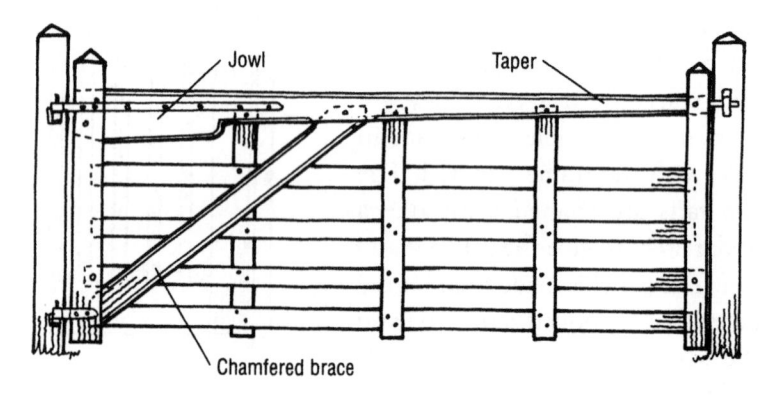

The Sussex field gate is the best of all field gates. Have a look at the project described later.

SWISS COTTAGE GATE

The Swiss cottage gate is a carriage gate with capped posts. The gate is framed, with two stiles and three rails, and the whole face of the gate is covered with sawed and fretwork balustrades in the Swiss tradition. The fretwork balustrades give a double pattern effect, with the shape of the individual balustrades and the patterns created by the spaces between neighboring members all adding to the total effect. This is an attractive option if you enjoy folk work and if you'd like to use either a scroll saw or a hand fretwork saw.

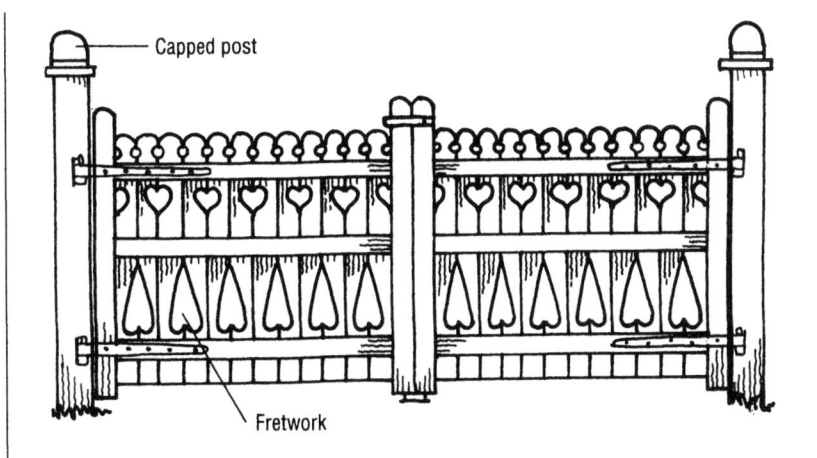

Known variously as a Swiss gate, a folk-art gate, a fretted picket gate, a Swedish gate, and many other names, these gates have the pickets close-butted so that the fretted holes create patterns.

TRUSSED HARR FIELD GATE

Although in many ways this is just another five-bar field gate, it is special in that the hinge stile, or harr post, has an integral bracket or truss to support the top rail. The truss is cut from the solid, which makes for an immensely strong and long-lasting gate. When old-timers made this gate in oak, they claimed that it literally lasted a lifetime—the full four score years. Although a trussed harr might not seem to be too much of a problem in terms of construction, you have to take into account that instead of being able to make the harr post from the usual 4-by-6-inch section, you need to start out with a massive 4-by-10-inch section. Most gates of this type were chopped out from oak boughs, with the bracket being cut from branch spurs.

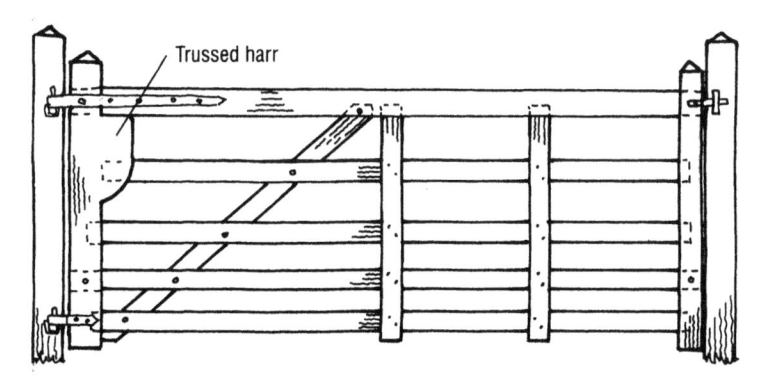

The best way to build a trussed harr gate is to search for a hefty burr or bough section that is flawed for most other purposes because of a knotty swelling on one side. You can cut the truss from the irregularity.

Generic Gate Types • 129

TURNED BALUSTRADE GATE

This framed gate has two stiles, three rails, and capped posts. The bottom is infilled with boards, and the top is filled with a line of turned spindles, or a balustrade. (Note that a single turned spindle is called a baluster, and a whole row of them is a balustrade.) The bottom is braced from corner to middle. If you want to make a gate for your driveway and you enjoy wood turning, this gate is a good option.

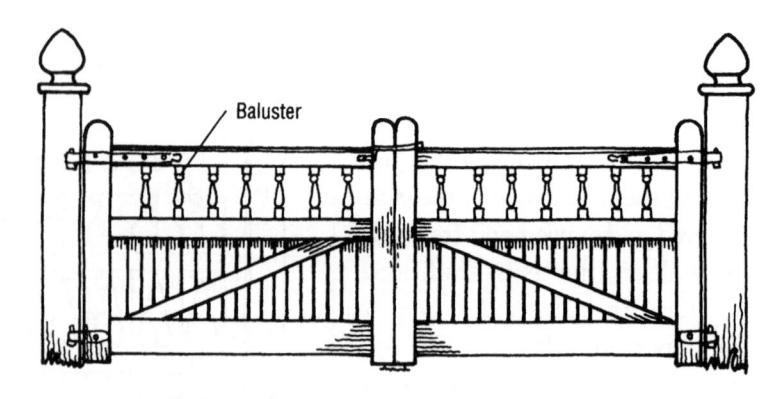

With this turned balustrade gate, the maker followed through with the design by turning finials for the posts.

TURNSTILE GATE

This framed gate has two stiles, three rails, two braces, and a number of vertical members set between the rails. What makes this gate special is that the entire frame is fixed with mortises and tenons. It is robust and obviously designed for heavy use.

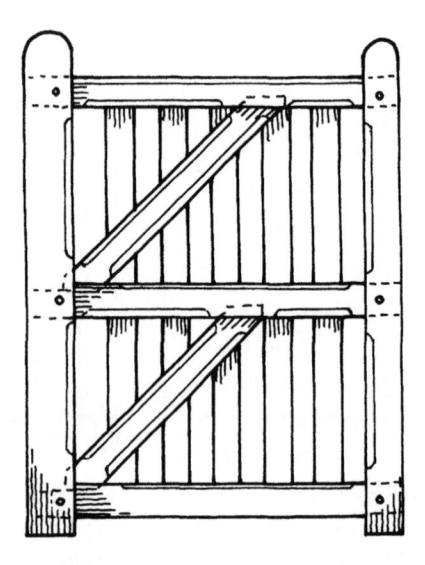

This turnstile gate is designed to be strong and long lasting. The through tenons at the ends of the rails, the stub tenons at the ends of the braces, and the chamfers on the edges that are most likely to weather are all details that ensure the longevity of the gate.

VICTORIAN GOTHIC GATE

The frame of this single carriage entrance gate with pyramidal top posts is made from two stiles and four rails with intermediate muntins. The bottom of the gate is boarded, and the top is open. Within the open top of the gate—in the six frames created by the pattern of the rails and the muntins—there are heavy Gothic frets. From left to right are a quatrefoil, four Norman lancet arches with top piercings, and another quatrefoil.

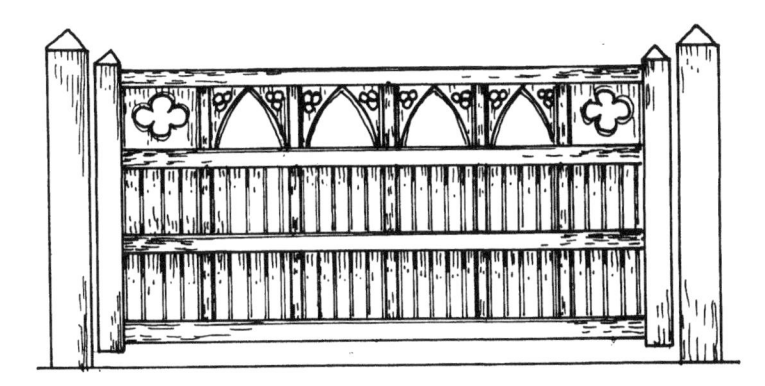

This Victorian Gothic gate looks the same from both sides. If you live in a Victorian vicarage, you have to make a gate for a church, or you have a yen for ecclesiastical woodwork, this is the gate for you.

WATTLE-HURDLE GATE

In this simple, traditional field gate made from coppiced hazel rods, the rods are split and woven between uprights to make a strong but lightweight gate. The two side poles—the stiles—are extended at top and bottom to become a pivot hinge. The bottom pole is spiked in the ground, and the top pole is tied to a post or a neighboring gate. To open the gate, the whole thing is untied and lifted.

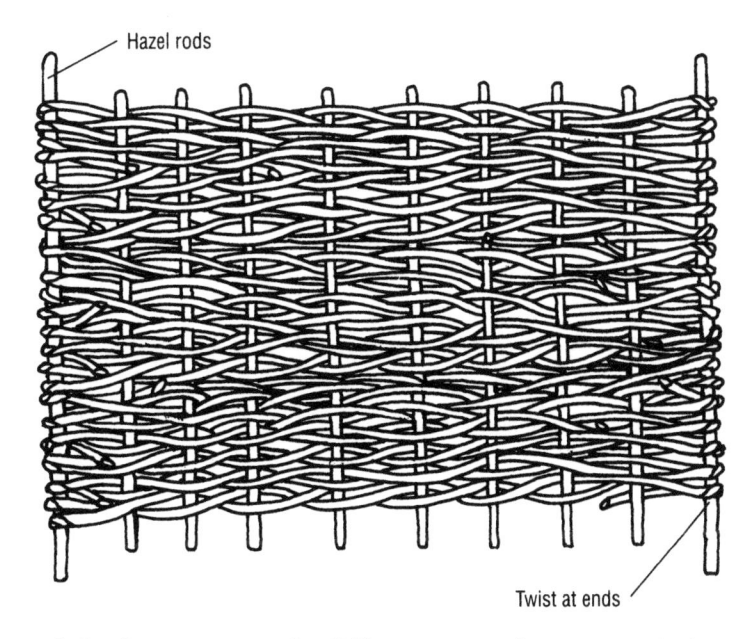

Wattle-hurdle gates were used in fields as temporary fences to contain sheep and the like and to keep the wind off vulnerable livestock such as newborn lambs. They are now more commonly being used in gardens as wind- and frost-breaks.

Generic Gate Types • 131

WEATHERBOARDED GATE

This framed gate has stiles, ledges or rails, and a brace. The rails are mortised into the stiles, and the middle and bottom rails are rabbeted to take the vertical weatherboarding. The tops of the stiles can be cut square or fancy with scrolls or rounding. If you like the overall shape of the gate but want something different from the vertical weatherboarding, you can have horizontal boarding or leave out the boarding altogether and have vertical bars and curved braces.

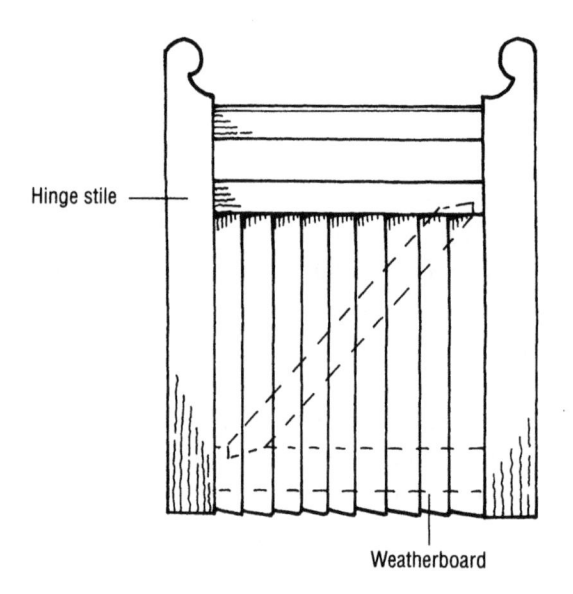

Weatherboarded gate, showing the front face. This is a plain and simple framed and braced gate.

WICKET GATE

This framed gate has two stiles, three rails, and a brace. The face of the gate is covered with a number of pickets or pales. Old gates of this type and character are generally made from two wood species—the frame of oak, and the pickets of pine. The rails are mortised through the stiles with bare-faced tenons, so that the surface of the frame is flush, and then the joints are fixed with oak pegs or pins. If you are looking to make a pretty, traditional gate for a cottage courtyard or garden, this is an attractive option.

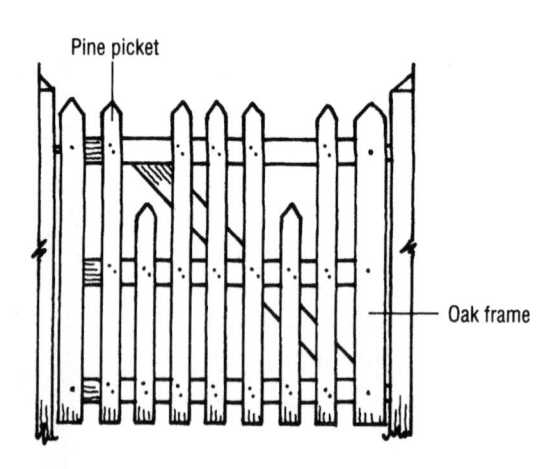

Apart from the rails being mortised into the stiles, all the other members of this wicket gate are secured with nails. The frame lasts just about forever, and the pickets can be easily replaced.

Making a Five-Bar Sussex Field Gate

Of all the five-bar field gates, the traditional Sussex gate is the best option. It's not only that all the sections make a particularly strong gate—a structure that will outlast most of us—but also that the sum of all the parts makes a uniquely beautiful and balanced form. But considering that the Sussex field gate evolved over several hundred years, it is hardly surprising that it's a winner.

If you study the working drawings, you will see that the Sussex gate is unusual. The top beam diminishes in depth as well as in width, and the sizes of the mortise and tenon joints vary from one end of the top beam to the other and from one rail to another. These features make it a tricky gate to build, but just think of the pleasure and the challenge!

If you decide to modify the design or use different wood or slightly different joints, you have to accept that the structure might be compromised. And if you decide to change the proportions—the length or height—although the gate might not lose any of its structural integrity, you might end up with a gate that is less than beautiful. So consider carefully before you make changes. If you have any doubts about the design and the order of work, make a scale drawing of all the details to clarify the sizes and joints.

TOOLS
- A pencil, a rule, a mortise gauge, a knife, and a square for marking.
- Three saws—a rip, a crosscut, and a fine-bladed bow saw.
- A good-size bench plane—a trying or smoothing plane—and a block plane.
- A large rasp.
- A paring chisel about 1½ inches wide.
- Two solid mortise chisels—one ¾ inch and the other 1¼ inches wide.
- A good-size mallet.
- A brace with a set of bits.

CUTTING LIST AND FIXTURES
Note that although the instructions call for well-seasoned oak throughout, you can cut costs by using oak for the stiles, beam, and brace and a lesser wood for the rails and slats. You need the following:
- A harr stile 4 feet 6 inches long, 5½ inches wide, and 3½ inches thick.
- A head or slam stile 4 feet 6 inches long, 3½ inches wide, and 2¼ inches thick.
- A beam or top rail 9 feet 6 inches long, 9½ inches wide and 3½ inches thick at the harr end, and 3½ inches wide and 2¼ inches thick at the small end.
- Four rails 9 feet 3 inches long, 3½ inches wide, and ¾ inch thick.
- A brace 5 feet 6 inches long, 5½ inches wide, and 1⅜ inches thick.
- Three slats 3 feet 6 inches long, 3½ inches wide, and ¾ inch thick.

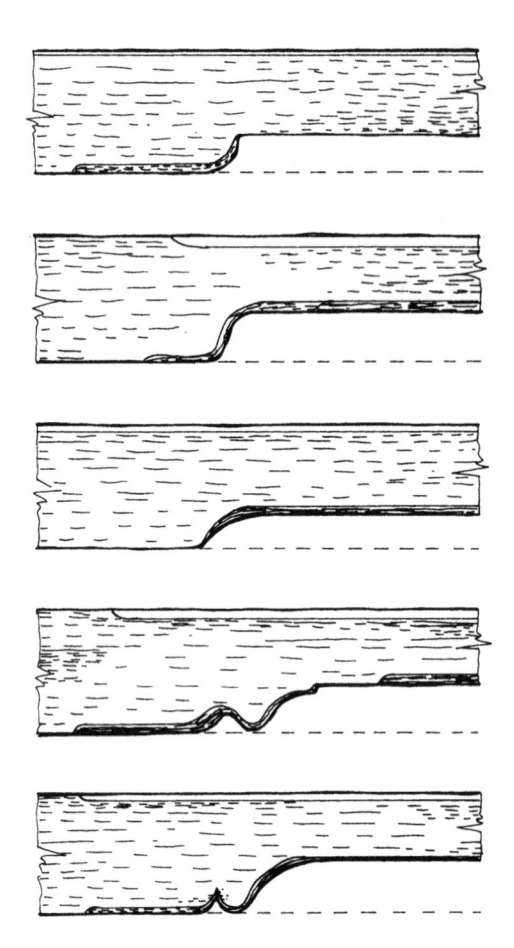

A selection of traditional jowl designs.

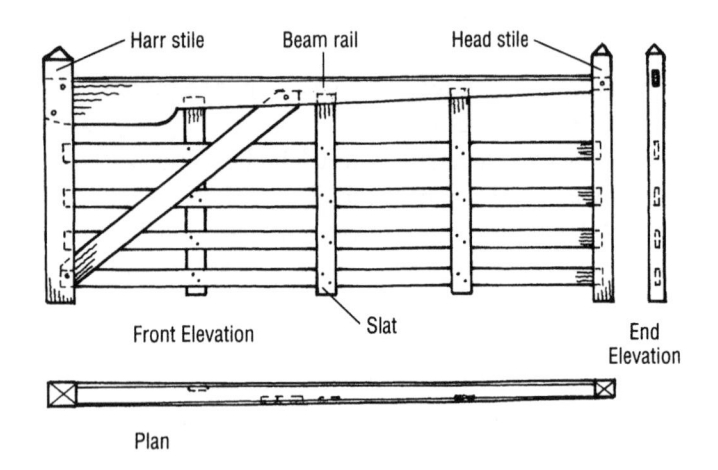

A traditional Sussex gate with a plain jowl, showing the front elevation, end view, and plan. Note how the top beam tapers in both views.

MAKING THE GATE
Preparing the Wood

1. After studying the working drawings and considering the various design, material, and tool options, give your wood one last inspection to make sure that it's free from flaws. Pay particular attention to the primary components: the two stiles, the beam rail, and the brace.

2. Saw the components slightly oversize, and then plane them back to the true width and thickness. Don't try for a perfect finish, as you would for an interior door; just make sure that the wood feels reasonably smooth to the touch. It's all pretty easy, apart from the top beam, which needs to be diminished from a 9½-by-3½-inch section at the harr end to a 3½-by-2½-inch section at the head end. The tapering occurs on all faces except the top or face edge.

3. Recheck the wood to make sure that the planing hasn't revealed a hidden problem, and then label each part. If you've done it right, you'll have eleven parts—all labeled, with the face side and face edge clearly marked.

Marking

1. Take the two stiles and set them so that the face sides are touching, then clamp them so that the face edges are uppermost and flush.

2. Being mindful that the beam rail mortises are

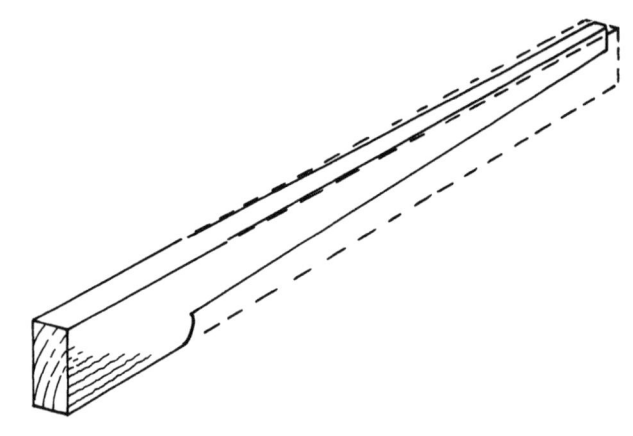

The feature that helps make a Sussex farm gate so special is the way the top beam tapers in both the front and the plan views.

134 • *Gates*

different sizes on the harr stile and on the head stile, take the square and mark the position of the top edge of the top beam rail and the mortises on the other four rails.

3. Remove the clamp. Then, because only the top beam is going to be worked with through mortises, continue the top lines across the face sides and across the other edge on both stiles.

4. Take the square and mark the top beam mortise on the stiles—9½ inches long on the harr, and 3½ inches on the head. Run these lines around the stiles so that they show on the back edge.

5. With the mortise gauge, gauge the top mortise on the harr stile to 1¾ inches, the top rail mortise on the head stile to 1 inch, and all the other mortises to ¾ inch. Have all the mortises centered on the stiles. This arrangement takes into account the fact that apart from the top beam, which is mortised through, all the other mortises will be stubbed.

6. With the square and pencil, measure about 1½ inches down from the top of both stiles and run the lines around on both faces and both edges. Run diagonals across the end grain—to establish the position of the end center point—and then use the square to run centerlines across both the width and the thickness of the sawed end.

Shaping the Pyramidal Stile Caps

1. Take the stiles one at a time, set them down flat on the bench, and get ready with the large paring chisel and the mallet. The best procedure for cutting the pyramid shape is to first slice in from both faces—so that you end up with a pitched roof shape—and then slice in from both edges. Some woodworkers prefer to slice one face at a time until the job is done. It sounds tricky, but you won't go too far wrong if you try to leave the end center point untouched and intact.

2. When you have removed the bulk of the waste with the chisel, clean up with the rasp. Work from the face or edge down to the end center point, so that you are cutting in the same direction as the run of the grain.

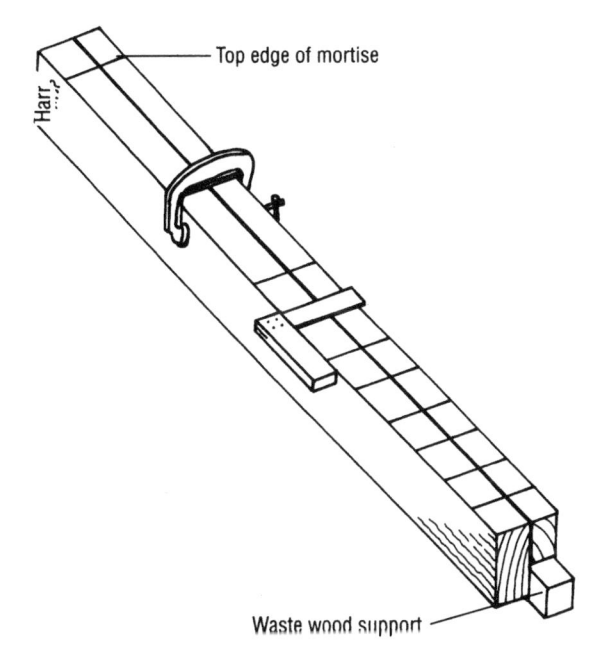

Clamp the stiles together, and use the square to lay out the mortises. Note that one stile needs to be supported with a block of waste to keep the whole works from tipping over.

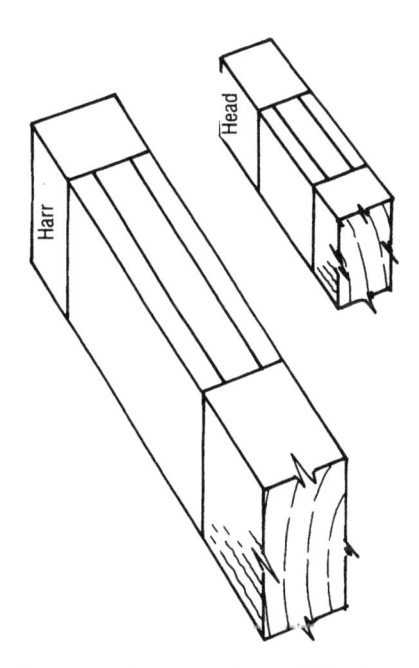

Gauge the two top beam mortises—1¾ inches wide on the harr stile, and 1 inch on the head.

Making a Five-Bar Sussex Field Gate • 135

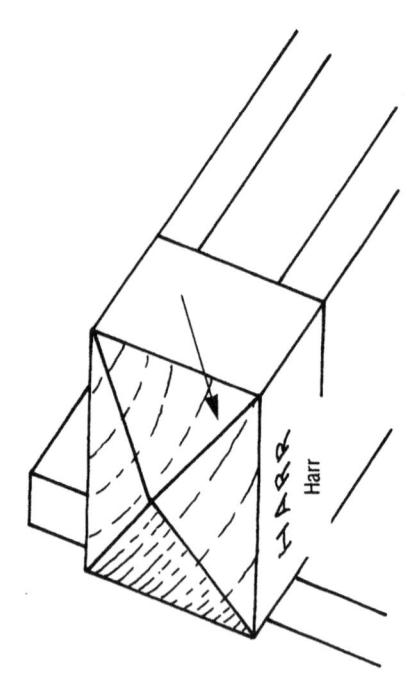

that it is marked off in four 2-foot 4½-inch lengths. If you have done it right, the beam should be set out from the harr end with six step-offs: 5½ inches, 2 feet 2¼ inches, 2 feet 2¼ inches, 2 feet 2¼ inches, 2 feet 2¼ inches, and 3½ inches. The top or face edge of the beam is at right angles to the ends, and the underside edge is diminished to run in a smooth line from one end to the other.

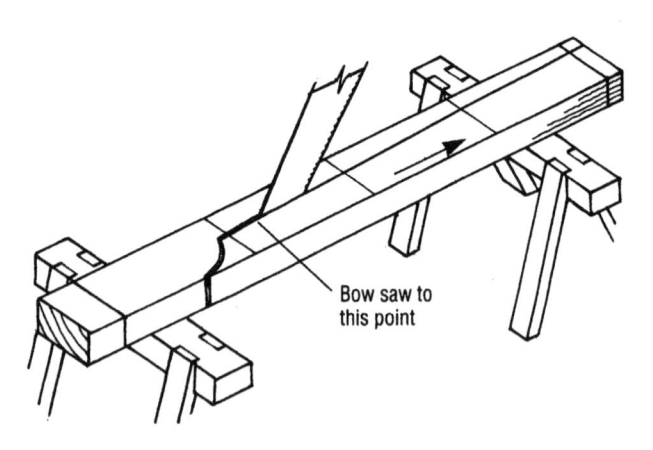

2. After studying the possibilities for the shape of the decorative jowl, use a pencil to draw the profile through to the face side. Figure to step up from the jowl to a beam width of about 6 to 7 inches at that point.

3. When you are happy with the shape, use a bow saw to cut the curves and a ripsaw to clear the long wedge of waste.

4. Plane the underside edge to a good finish, and use a mallet and chisel to tidy up the curves.

Chopping the Mortises on the Stiles

1. Check and double-check the position of the mortises. Being mindful that only the top beam mortises are to be cut through, position the stiles face edge up on the bench, and set to work with the mallet and the appropriate-width mortise chisel.

2. The mortising procedure is as follows:

• Cut a small V-shaped piece in the center of the mortise.

• Set the chisel in the V so that the bevel is facing away from you, and then work toward your body, cutting a succession of deeper cuts at ¼-inch intervals.

• When you get within about ¼ inch of the end of the mortise, lever out the chips of waste, and then turn the chisel around and work toward the other end.

3. For the stub or stump mortises, you stop at a depth of about 2 inches, whereas with the through mortises, you stop when the mortise is about halfway through, and then turn the stile around and chop in from the other side. Note that the through mortise on the harr stile needs to be diminished.

Slice down from side to center, so that you are always cutting with the grain.

Shaping the Beam Rail

1. Take the sawed and planed top beam, and set it face side up on the bench. Use the pencil, rule, and square to mark off the allowances at each end for the depth of the through mortises—5½ inches at the harr end, and 3½ inches at the head end. Between these two points, you should be left with a length of 8 feet 9 inches. Quarter this length, so

Fret the shape of the jowl with the bow saw, and then rip through.

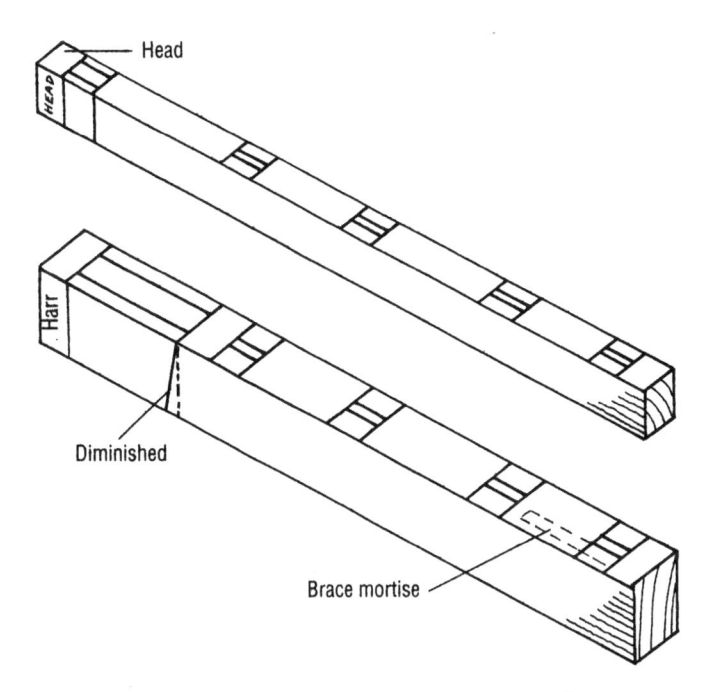

Marking the mortises is tricky, so spend time making sure that they are right.

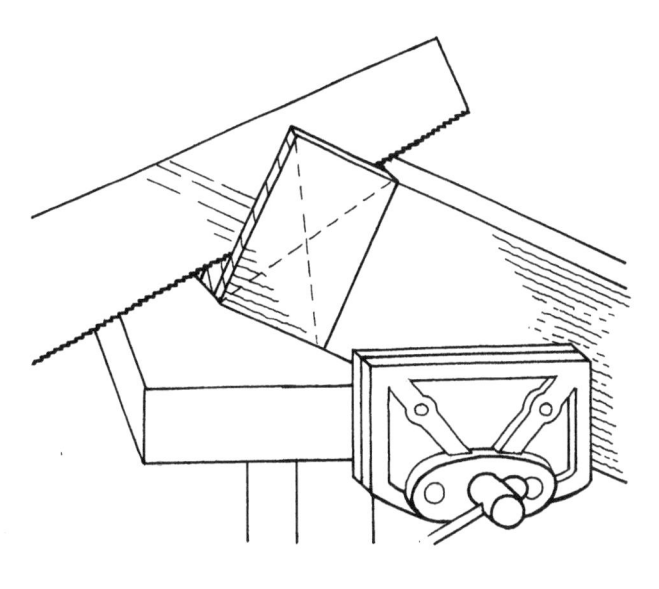

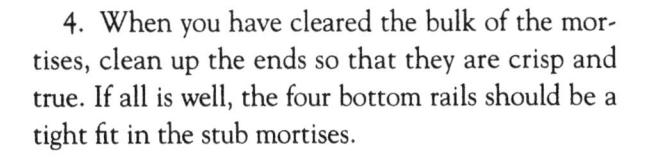

Make three cuts down to the shoulder line.

4. When you have cleared the bulk of the mortises, clean up the ends so that they are crisp and true. If all is well, the four bottom rails should be a tight fit in the stub mortises.

Cutting the Tenons on the Top Beam

1. Having set the lengths of the tenons out on the ends of the top beam—5½ inches at the harr end, and 3½ inches at the head end—use the square and knife to mark out the shoulders, and the mortise gauge to scribe the thickness. The harr tenon is 1¾ inches thick, and the head tenon 1 inch.

2. The procedure for cutting the tenons is as follows:

• Set the workpiece at an angle in the vise—so that you can see the edge of the tenon—and run cuts down to the shoulder line.

• Turn the workpiece around in the vise and repeat the cuts for the other edge of the tenon.

• Having established the width of the tenon on both edges, run cuts straight in from the ends to cut the remaining peak down to the shoulder line.

• Set the workpiece down flat on the bench, and saw down to the waste side of the shoulder line to remove the waste from the cheek. Do this on both faces of the tenon.

3. Finally, use a chisel to reduce and cut back the underside edge of the harr tenon, so that it slides into the harr stile.

Cutting and Fitting the Brace

1. When you have cut the stile mortises and the top beam tenons, have a trial fitting of the top beam and the other four rails in the stiles. Set the whole works square.

2. Set the brace across from the bottom of the harr stile—to just before the middle of the underside of the top beam—and mark the position and shape of the stub or stump tenons. Because the tenons are bare-faced, the shoulder lines need only be marked on the face side.

3. Cut the tenons on the ends of the brace in much the same way as already described.

4. Use the tenons on the brace to fix the position of the mortises on the underside of the top beam and the side edge of the bottom of the harr

Making a Five-Bar Sussex Field Gate • 137

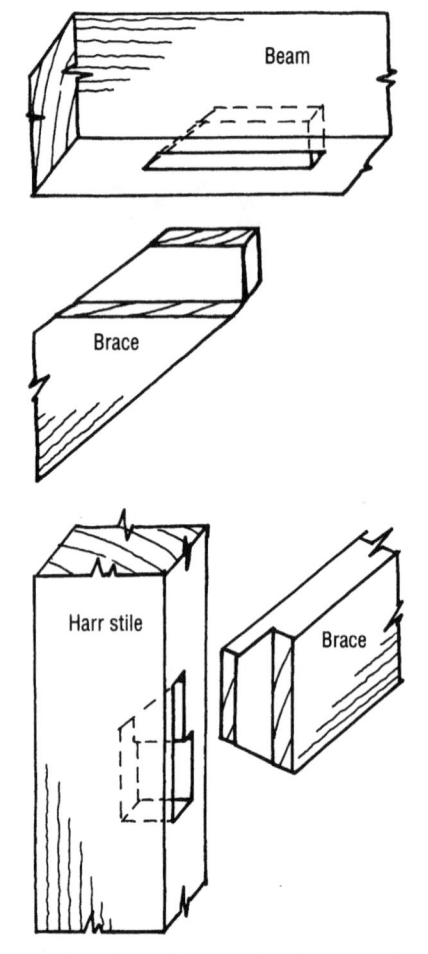

The brace needs to be stub-tenoned—the top end into the underside of the beam, and the bottom end into the bottom of the harr stile. If you've got it right, the bottom end of the brace and the harr end of the bottom rail will match up to share the same mortise.

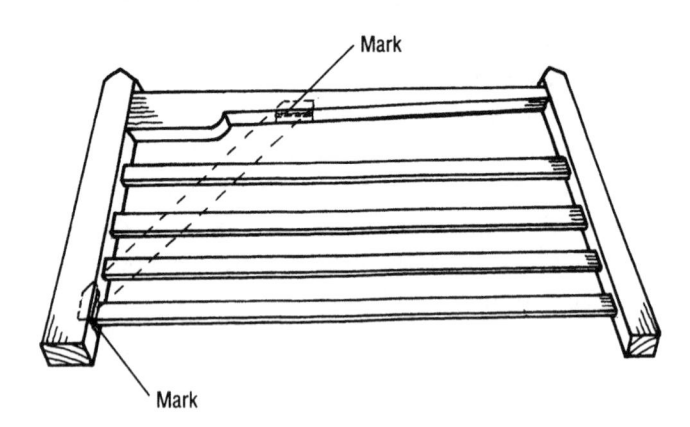

stile. Cut them as already described. If you have done it right, the two bottom mortises on the harr stile—the mortises for the bottom rail and the brace—will run into each other.

5. Finally, have a trial fitting of the brace, and see how it all comes together. If all is correct, the top beam and the harr stile will be locked at right angles, in such a way that the side faces of all three members—the harr, the beam, and the brace—are flush with one another.

Assembling

1. When you have cut the brace and had a trial fitting of all five rails between the stiles, set the whole works flat on the floor so that the harr stile is at the left. Now take the three slats and set them in place on the gate. If the three slats are numbered 1, 2, and 3 from left to right, have number 1 set between the rails and the floor, and numbers 2 and 3 set on top of the rails. The slats need to be spaced so that the gate is quartered across its width.

2. When you are happy with the spacing, mark the position of the stub mortises on the underside edge of the top beam, and cut them in the same way as already described.

3. Use a small plane to chamfer the stiles, the beam, and the brace.

4. Finally, draw the whole gate together with rope clamps, drill holes through all the joints and

Set the rails in place, and carefully mark the position of the brace mortises.

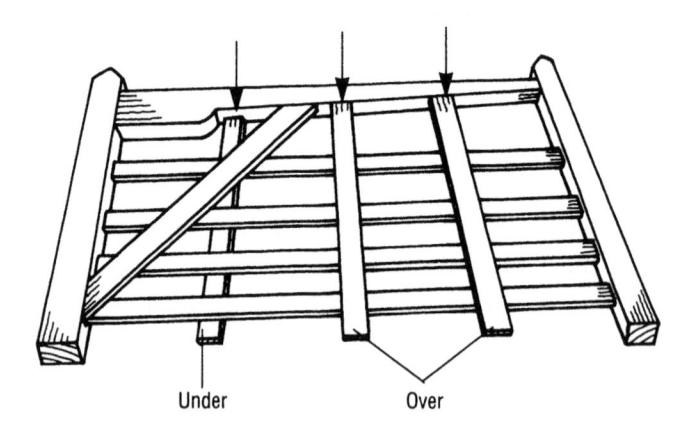

Set out the slats, and mark the position of the mortises on the underside edge of the top rail.

138 • Gates

through all overlapping members, and fix with either large clinched wrought nails or galvanized nuts and bolts.

AFTERTHOUGHTS AND OPTIONS

- When you are choosing wood, try to avoid large knots. This is especially important with the stiles and the top rails.
- Search for well-seasoned wood. Avoid wood that looks spongy or discolored.
- If you are unable to set the beam in the vise—because the vise is too small, awkwardly placed, or whatever—clamp the beam to a bench or across a couple of sawhorses, sit astride it, and work from that position.
- In the context of making farm gates, a mortise chisel is a massively built framing-type chisel with a strong, thick section blade.

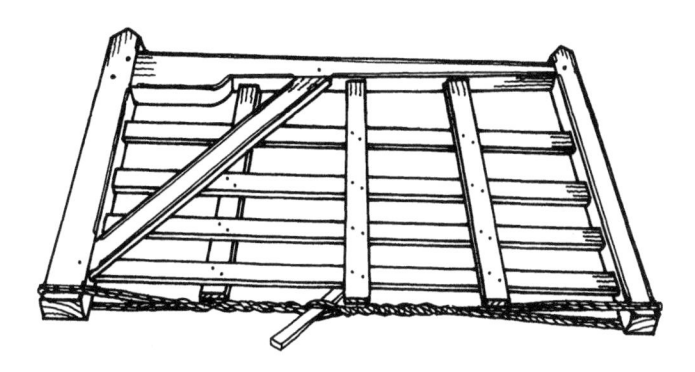

When you're clamping up, make sure that you use a couple turns of good strong rope and a stout ash stick.

- If you are going to make a top-class gate in an expensive wood such as oak and are worried about technique or the position of the joints, make a prototype in cheap wood first.

Making a Framed Oak Garden Gate

There is nothing quite so evocative and welcoming in the context of an ordinary country cottage as a plain and simple oak gate. Such gates are the perfect coming together of form, function, decoration, and design. They look good, they keep kids in and cattle out, they are good for swinging on, they never need painting, they don't jar with their surroundings, and they will still be standing generations later. So if you are looking to build a gate for a small, unpretentious country home, this may be the gate for you.

The following project describes how to make an oak gate 45 inches high by 33 inches wide with two stiles, three rails or ledges, a brace, and paneling made from sawed oak feather-edged weatherboarding. You can even modify the gate somewhat without worry, as long as you keep everything simple and low-key. You might want to make it from a lesser wood and paint it a traditional color such as white or green, but don't be tempted to deck it out with fussy brass fittings. And stay away from yacht varnish; once the gloss wears off, it soon begins to look old and worn.

TOOLS
- Three saws—a rip, a crosscut, and a tenon.
- Three planes—a smoothing, a bench rabbet, and a block.
- A pencil, rule, mortise gauge, knife, and square.
- A mortise chisel approximately one-third the width of the frame thickness—a ½-inch chisel would be fine.
- A brace with a selection of bits.
- A mallet and a hammer.

CUTTING LIST AND FIXTURES
We have chosen to use American oak throughout and have allowed a generous amount of extra length for cutting waste. You need the following:
- Two stiles 3 feet 10 inches long, 3¾ inches wide, and 1⅝ inches thick.
- The top rail at 2 feet 10 inches long, 3¼ inches wide, and 1⅝ inches thick.
- The middle rail at 2 feet 10 inches long, 4¼ inches wide, and 1⅝ inches thick.
- The bottom rail at 2 feet 10 inches long, 4¼ inches wide, and 1 inch thick.
- A brace 2 feet 9 inches long, 4¼ inches wide, and ⅞ inch thick.
- Two 37-inch pieces of 1-inch square section for the side fillets.
- Ten feather-edged weatherboards about 2 feet 6 inches long, 4 inches wide, and ⅝ inch thick—this allows for a generous amount of sorting and cutting waste.
- A generous amount of scrap wood for fixing pegs.
- A quantity of 1-inch nails for fixing the weatherboards—copper is preferable.
- A handful of 2-inch galvanized nails for fixing the side fillets.

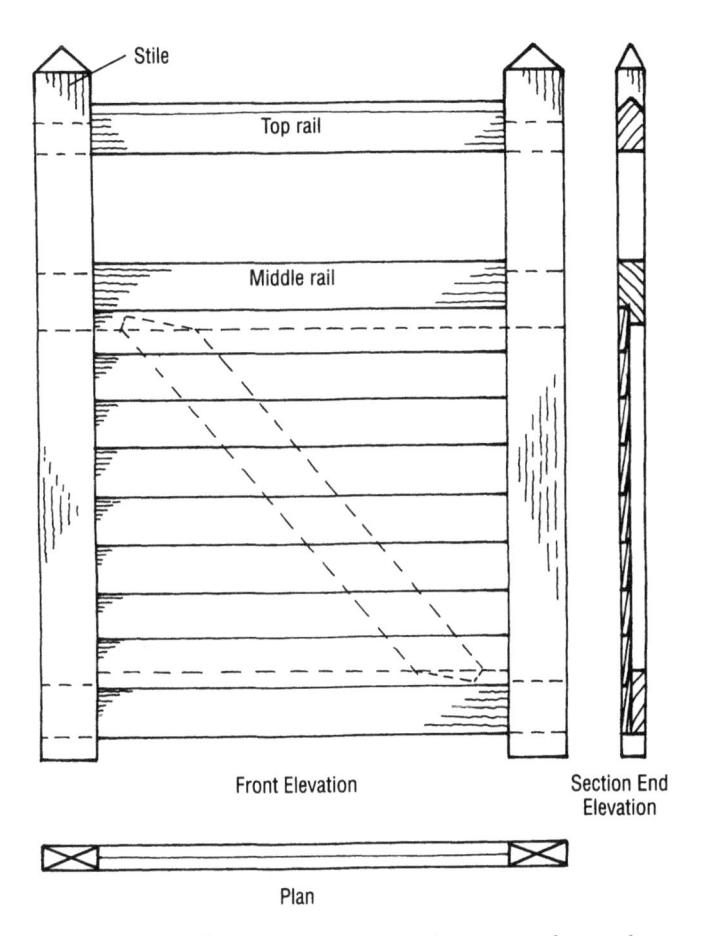

A traditional front entrance gate—front view, plan, and side end elevation.

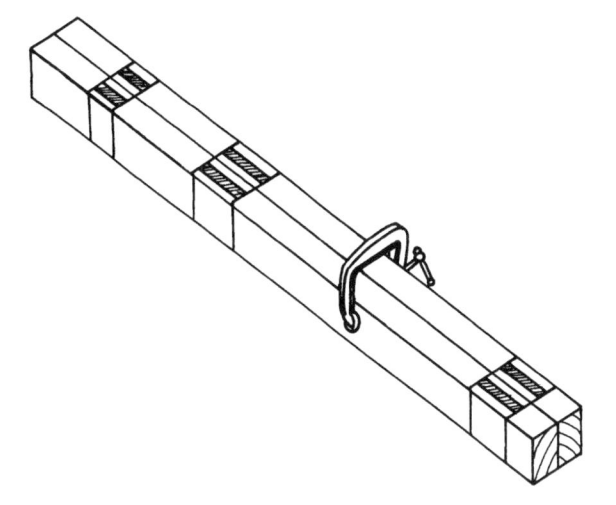

Although there is no reason why you can't have the rails set at a different spacing, it is vital that, from stile to stile, the mortises are laid out identically.

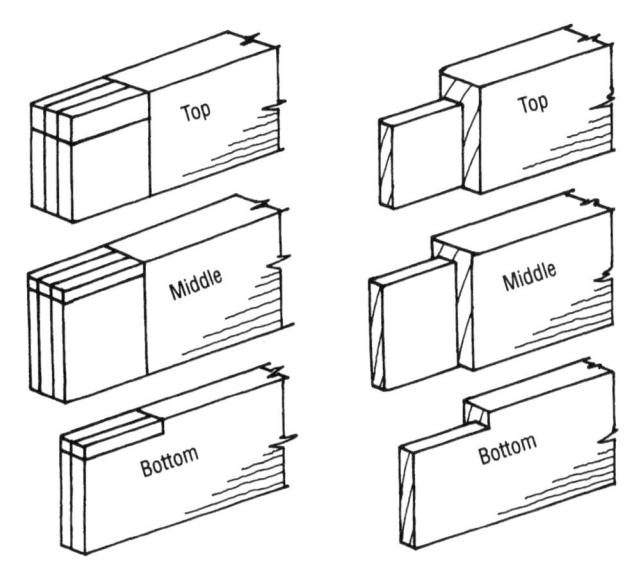

Left: *The rails marked out with the tenons.*
Right: *The completed tenons.*

MAKING THE GATE
Preparing and Marking

1. After studying the designs and carefully selecting your wood, plane the two stiles to a finished size of 1⅜ inches thick and 3½ inches wide.

2. Mark out the position of the three mortises. Have all the mortises ½ inch wide, centered, and running right through the stile. The top mortise will be 2¼ inches long and ½ inch wide, and the middle and bottom ones will both be 3¾ inches long and ½ inch wide. Working from the top of the stile, the spacing between mortises is as follows: 6 inches at the top for waste and shaping, 2¼ inches for the top mortise, 4 to 6 inches for a space between the top rail and the middle rail, 3¾ inches for the middle mortise, about 22 inches between the middle rail and the bottom rail, and 3¾ inches for the bottom mortise. You should be left with about 2 to 3 inches below the bottom rail.

3. When you are happy with the spacing of the rails and the layout of the mortises, take a pencil and square and run the mortise lines around to both edges of the stile. Gauge the mortises at ½ inch.

Making a Framed Oak Garden Gate • 141

4. Plane the rails to size, cut them to a finished length of 33 inches, and then use the square and gauge to lay out the size and position of the tenons. With the tenons all being 3½ inches long and ½ inch thick, the top rail tenon will be 2¼ inches wide, and the middle and bottom rail tenons will both be 3¾ inches wide. Note that the bottom rail tenon is bare-faced.

Cutting the Mortises

1. When you have checked and double-checked the position of the mortises, clamp the two stiles side by side on the bench and set to work chopping out the waste. Although this procedure can be done in any number of ways, it's best to cut a V-notch at the center and then back up, making deeper and deeper cuts and stopping just short of

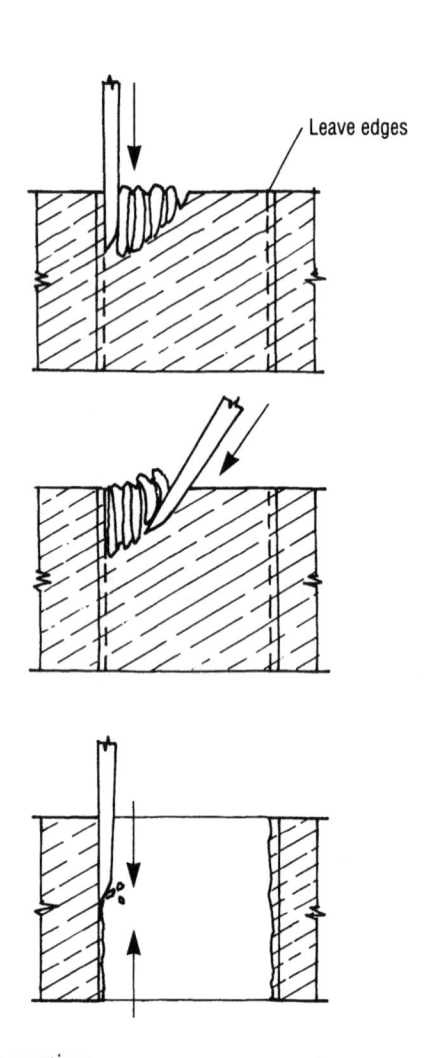

Cutting the mortises.
Top: *Back up from the V-cut—making deeper and deeper cuts as you go—to stop just short of the end of the squared line.*
Middle: *Lever out the waste chips, and then repeat the procedure for the other end and for subsequent levels.*
Bottom: *When you have chopped through from both sides, pare the ends to the squared line.*

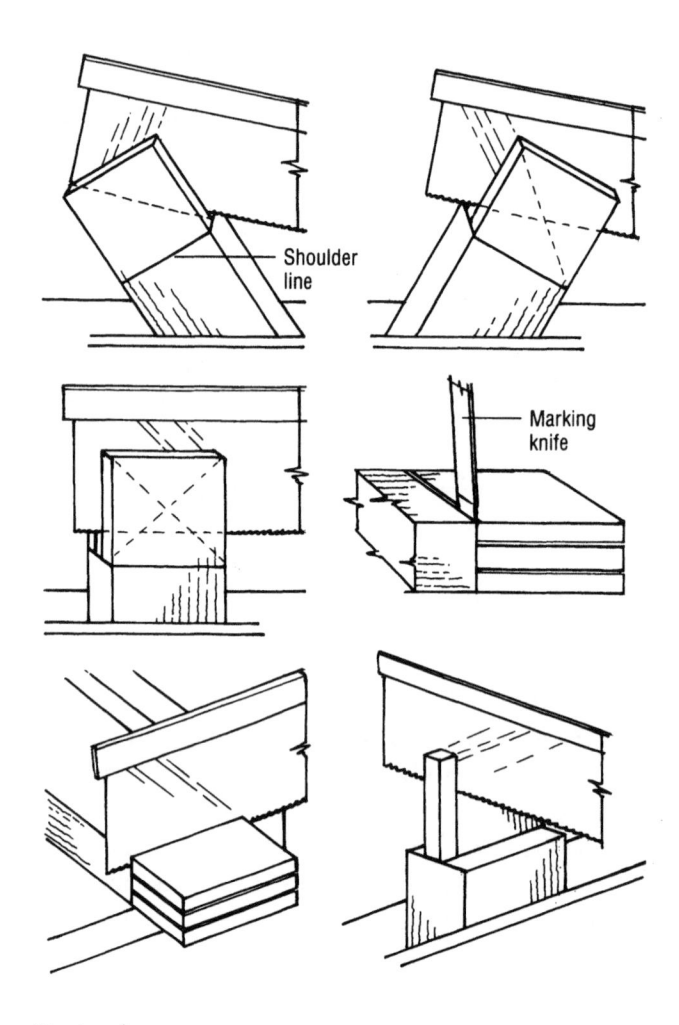

Cutting the tenons.
Top left: *Saw down to the shoulder line.*
Top right: *Tilt the wood in the other direction and saw down to the other shoulder line.*
Middle left: *Saw straight down to the shoulder line.*
Middle right: *Score the line of the shoulder.*
Bottom left: *Saw away the waste to form the cheek and the shoulder.*
Bottom right: *Cut away the small strip of waste to form the third shoulder.*

142 • *Gates*

the ends. Once this is done, lever out the waste, turn the chisel around, and chop out the other end. When you have cut the mortises halfway through, turn the stiles over and chop them through from the other side.

2. Finally, with all waste removed, make downward thrusting cuts into the mortises to remove the rounded-over waste at the ends.

Cutting the Tenons

1. Now use the tenon saw to cut away the waste faces of the tenons as follows:
- Set the rail at an angle in the vise and saw down to the shoulder line.
- Turn the rail over in the vise—still at an angle—and saw down at an angle on the other side.
- Set the workpiece upright in the vise and use the two kerfs as a guide to saw down to the shoulder mark.
- Set the workpiece flat down on the bench and use a knife to make a V-cut to mark the shoulder line.
- Set the rail against a bench hook and saw down to the cheek to remove the waste.

2. Have a trial fitting and see how the frame comes together. If necessary, use a chisel to ease the joints.

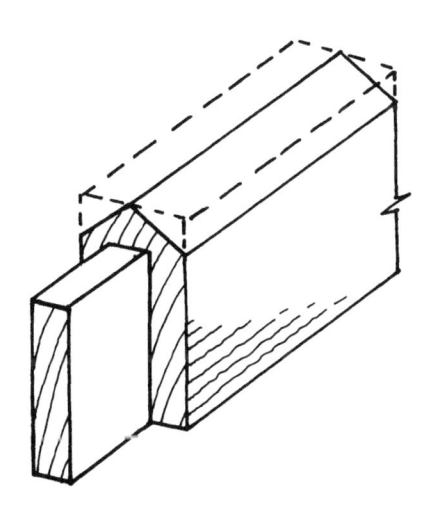

Cutting the Chamfers and the Rabbet

1. Once the mortise and tenon joints are nicely fitted, take the plane and run a double chamfer on the top rail and slight bevels on the other two rails.

2. Use the bench rabbet plane to run a rabbet ½ inch deep and 1 inch wide on the underside of the middle rail. If you have done it right, the top and middle rails will both show a width of 3 inches when the weatherboarding is in place.

Fitting the Brace

1. Set the brace in place across the front of the gate, allowing about 1 inch between the ends and the stile, and then pencil-mark the position where it crosses the rails.

2. Use the saw to cut bare-faced laps or tongues on the end brace, and then use the chisel to notch the rails to fit. If you have done it right, the face of the rabbet on the underside of the middle rail and the face of the bottom rail should run through

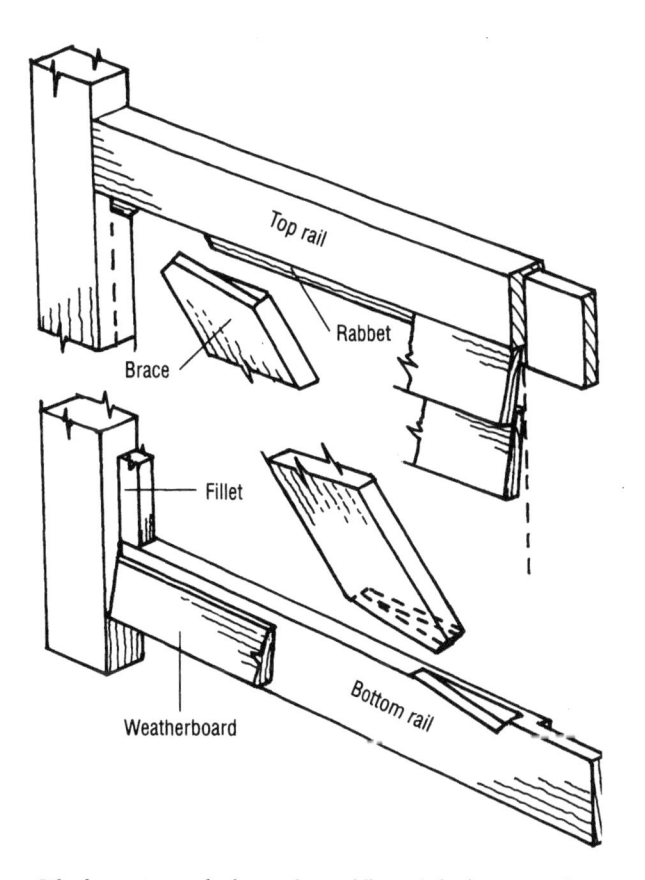

Run a chamfer on each side of the top of the rail to create a double-pitch section.

The brace is notched into the middle and the bottom rails.

Making a Framed Oak Garden Gate • 143

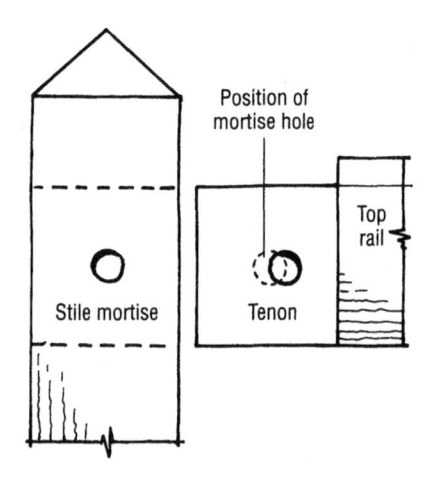

The draw-boring technique relies on the holes being out of alignment, so that the hole through the tenon is set slightly off-center and nearer to the shoulder.

flush with the brace, providing a bed for the weatherboarding.

Assembling and Finishing

1. Use the saw and plane to pretty up the top of the stiles. You can simply round them over at the corners, cut pyramids, put chamfers on at the sides, cut stylized scrolls, or do whatever suits your fancy.

2. Have a trial fitting to make sure that the rails are correctly placed and everything is square and true. Then set to work draw-boring and pegging the mortise and tenon joints as follows:

- Bore a hole through the center of the mortise—in one side and out the other.
- Fit the tenon in the mortise and mark the center of the hole through to the cheek of the tenon.
- Pull the joint apart and run an off-center hole through the tenon, making the hole slightly nearer the shoulder.
- Put the joint together and run a steel draw-boring pin through the hole to pull the stiles hard up against the shoulders. Then remove the steel pin and follow up with whittled oak pegs.

3. Finally, when the frame is all tightly drawn together with the oak pegs, plane the fillets to size and nail the ⅞-by-⅞-inch fillets to the side of the stiles so that the front face is flush with the rabbets and the brace. Then cut the weatherboards to length and nail them in place.

AFTERTHOUGHTS AND OPTIONS

- Although we particularly like oak, you can use another suitable hardwood or even painted softwood. However, a gate of this character is better made from a native hardwood.
- Never use an endangered species such as mahogany. It's politically incorrect, it looks out of place, it's expensive, and worst of all, it weathers badly.
- If you decide to paint the gate, use a traditional color such as black, cream, green, or maroon. And it's best to paint the parts before you put the gate together—even inside the joints.
- Be mindful that oak and iron are a bad mix—the iron swiftly corrodes, and the oak stains black. Avoid steel and use copper or galvanized metal.
- The draw-bored pegs are perfectly adequate for holding the joints. There is no need to use glue or bolts.
- Although here the weatherboarding has been fixed horizontally, there's no reason it can't be fixed vertically or even diagonally.

Hanging a Field Gate

Of all the gates, a wide field gate is the most difficult to hang, running the gamut of problems. Whereas little domestic gates can be hung in much the same way as braced and ledged doors—with strap hinges and the like—field gates have long members that lever down on the hanging post, to the extent that the post may lean over, making the gate sag and drag on the ground. The effect of a badly hung field gate—or any gate—is that it's a nuisance to open and close.

There are four factors to consider when you are hanging a field gate: the size and shape of the hanging post (the post to which the gate is hinged), the amount of post belowground, the way the post is fixed and held in the ground, and the way the hinges are constructed. Most important of all is the amount of post in the ground in relation to the height of the post and the width of the gate.

Traditionally, in England, the Sussex-type field gate was hung on an oak post—a post about 12 inches square was the norm—with at least half the length of the post belowground and the belowground half being left untrimmed. All this adds up to a hanging post that is bottom-heavy—a bit like a yacht with a heavy iron keel. Old-timers claim that to set the post in the ground in concrete is the worst thing you can do. They believe that concrete encourages rot, and the lack of give results in the post snapping off at the base. The best method is to dig a deep hole that widens at the base, set the post in place, and ram the hole full of earth and stone. If you do it properly, such a hole will require more filling than was originally dug out. The proof that the ramming method is successful is that many posts erected well before the Second World War are still going strong.

As to the hinges and catches, the most satisfactory arrangement is to have a hinge hook bolted through the hanging post, large strap hinges with side wings bolted to the harr post, large strap hinges with side wings bolted to the harr post and through to the top rail, a U-shaped catch bolted to the slamming post, and a sprung fastening bolted to the head post. The advantage of the U-shaped catch is that besides being relatively easy to manage, the weight of the gate on the hanging post is relieved when the gate is closed by being supported on the slamming catch. The hinges have been selected so that the gate is self-closing. The bottom hook is set out slightly so that the toe is raised. Although such an arrangement results in the gate looking a bit askew when it is closed, the benefit is that the gate always swings closed.

TOOLS

- Earthmoving tools such as a spade and a rammer or tamper.
- A crosscut saw.
- A brace with a good selection of bits.
- A pencil, rule, tape measure, and square.
- A hammer.
- A long spirit level or a plumb line.
- Spanners to fit the various bolts.
- A pair of heavy leather working gloves.

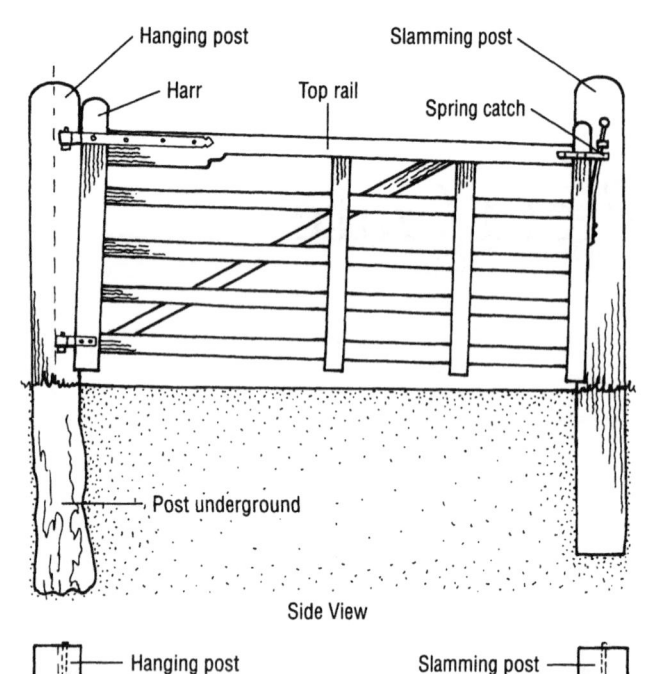

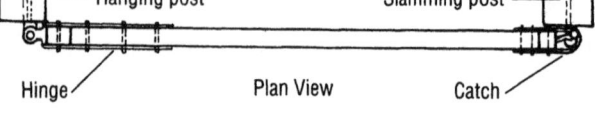

Front view of the gate with a section through the ground and a plan view to show the relationship between the gate and the posts.

CUTTING LIST AND FIXTURES

- Two 12-by-12-inch square section oak posts—one about 9 feet long for the hanging post, and the other 8 feet long for the slamming post. If the bottoms of the posts are rough as they come from the tree—unsawed and rounded—so much the better.
- A top hinge for a field gate, meaning a strap hinge about 2 inches wide, ⅜ to ½ inch thick, and about 2 to 3 feet long.
- A bottom hinge about 2 inches wide, ⅜ to ½ inch thick, and 4 inches long.
- Nuts, bolts, and washers to fit your chosen hinges.
- Two hinge hooks—with threaded shanks long enough to go through the gateposts—with nuts and washers to fit.
- A sprung catch or fastening set, with bolts to fit.
- A good supply of scrap wood for temporary support.
- Crushed brick and rubble to go into the holes—soft brick, mortar, and the like, not broken concrete.

HANGING THE GATE
Measuring and Fitting the Posts

1. Before you start digging holes, make sure that the hinge hooks and the catches are fitted on the slamming face of the posts. That is, when you are on the side on which the gate opens and are looking at the posts, the hinge hooks and the catches should all be on the side of the posts facing you.

2. After you have chosen the site and have a clear understanding of precisely where you want the gate to be, start by digging the hole for the hanging post. Aim for a hole that is a little bigger than the widest part of the post at the mouth, getting wider as it gets deeper. For example if the gatepost is 9 feet long, there are 6 inches between the top of the stile and the top of the gatepost, and the gate stile measures 4 feet 3 inches from top to bottom, with a minimum 3-inch clearance between the bottom of the stile and the ground, then you need to dig the hole about 4 feet deep.

> **SPECIAL TIP**
> Hanging field gates is really a two-person task. You'll need help lifting and maneuvering the posts and the gate as well as someone to give you a yea or a nay when you are truing up the posts.

3. Once the hole has been dug, put a brick or rock in the bottom, then take the hanging post and set it vertically in the hole. Test it with a spirit level or plumb line, and arrange the hinge face so that it's facing the opening side. Either brace and support the post with a couple of battens or get a friend to hold it upright.

4. When you are happy with the arrangement, drop a small amount of earth and crushed brick or rubble into the hole and ram it home with the tamper. Continue putting earth and bits of brick into the hole and ramming it home, systematically working around the post until the hole is topped

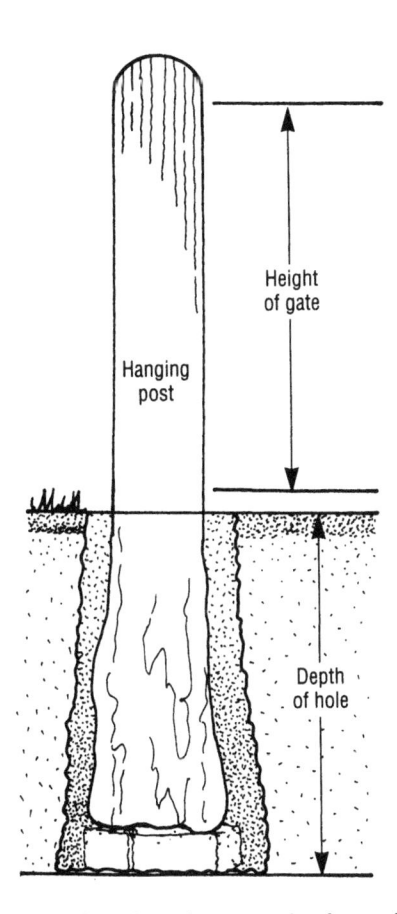

Set the post in the hole at the correct height, so that the gate clears the ground.

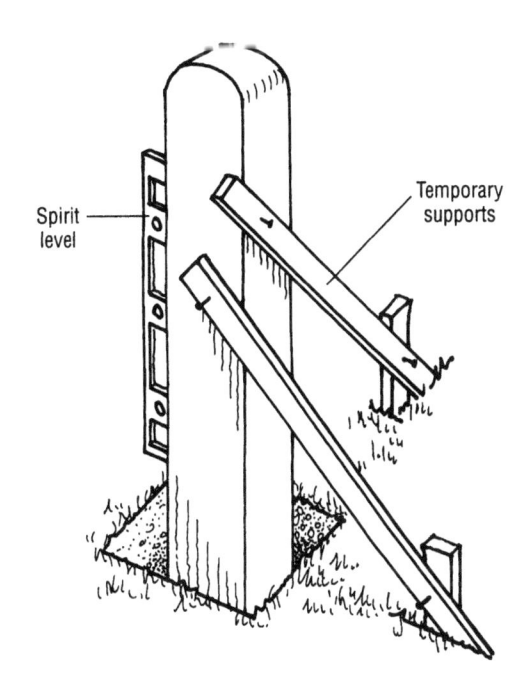

Set the post upright, and fix temporary braces to hold it in that position. Use a spirit level or plumb bob to check for vertical alignment.

up. The secret is to ram hard, to keep testing whether the post is true, and not to get to the point where you need to heave the post back to true. Old-timers always replaced the turf to prevent the earth from washing away and forming a sump.

5. With the hanging post firmly planted, measure the length of the gate—ours is 9 feet 6 inches—and mark off this distance for the slamming post. Have the center points of the two posts precisely 9 feet 6 inches apart.

6. When you have checked and double-checked the measurements, plant the slamming post in exactly the same way as already described, only this time around, set the post in a shallower hole about 3 feet deep.

Fixing the Hinges

1. Set the gate on the ground or across a couple of sawhorses with the best face or opening face up.

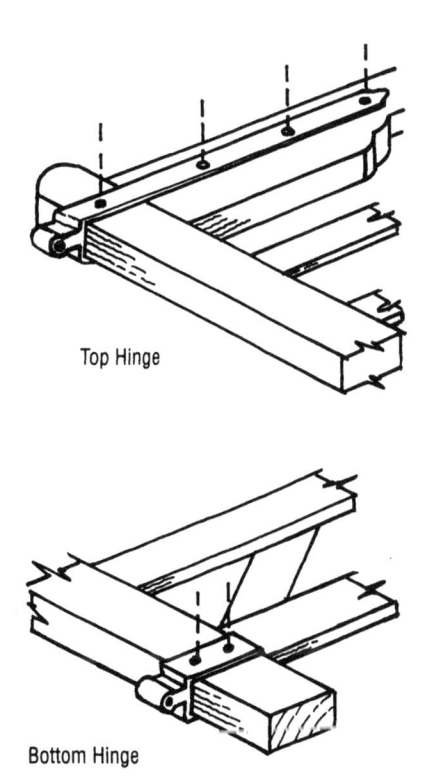

Set the hinges in place and fix them with through bolts. Note that the wings of the top hinge give the gate added strength by being bolted to both the stile and the top beam.

Hanging a Field Gate • 147

Set the two hinges in place—one toward the top of the harr stile and centered on the top beam, and the other toward the bottom of the harr and centered on the bottom rail. Mark the position.

2. When you are happy with the placement, bore holes through and fix with bolts. Make sure that you put washers between the wood and the nuts.

Fixing the Hinge Hooks

1. With the two posts firmly planted in the ground and the hinges bolted to the gate, take a pencil, rule, and spirit level and draw in clear centerlines on the opening or slamming face of both posts. If you have done it right, the two centerlines will be vertical and parallel to each other and 9 feet 6 inches apart.

2. Taking into account that the bottom of the harr stile needs to be lifted about 3 inches clear of the ground, draw the position of the two hinge hooks on the hanging post centerline. Then shift the position of the top hook about 1 inch to the right of the centerline.

3. Bore a hole through the post for the bottom hook—dead on the centerline—and bolt it in place.

4. Sit the gate on the bottom hook, with the gate closed flat against both posts, prop the toe up so that it is about 1 inch clear of the ground, and check that the placement of the top hook is correct. If you have done it right, the gate should be tilted down toward the toe, and the top hook should be a little to the right of the post's centerline.

5. When you are satisfied that all is correct, bolt the top hook through in the same way as already described, clinch the nuts tight, and then drop the gate in place. If necessary, adjust the angle of the hooks by slackening off the nuts and swiveling them around slightly.

6. Now for the big test: If you have done it right, when you open the gate, the toe will rise up until the gate is fully opened. If you release the gate at any point in its opening arc, it will always fall back into the closed position.

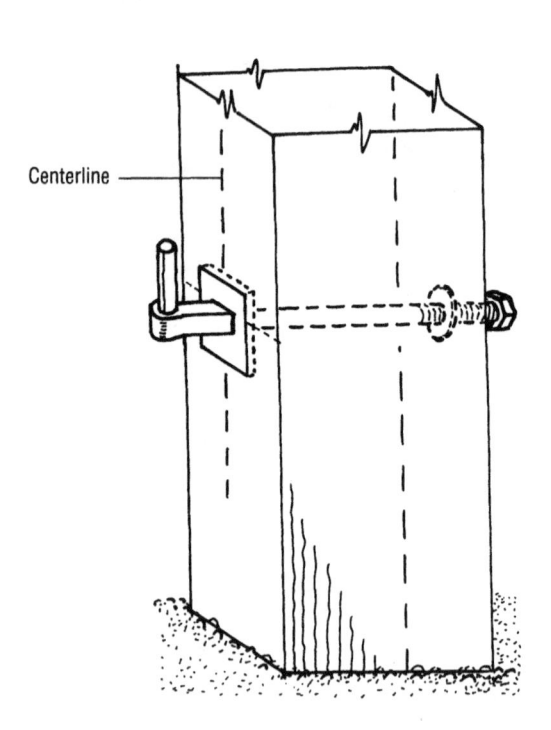

Set the top hinge hook slightly to the right of the centerline and secure with a washer and a nut.

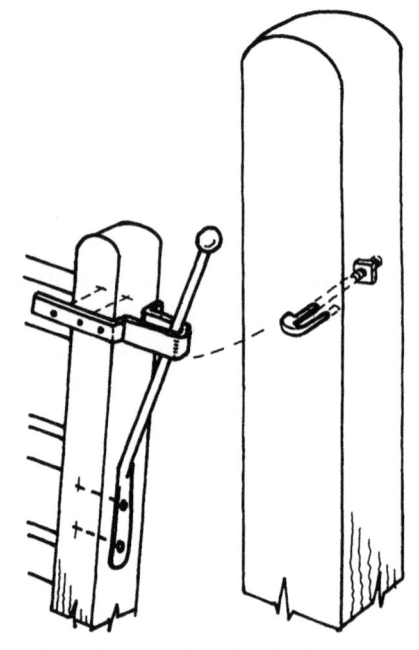

Fix the catch so that it is aligned with the spring arm—so that it positively engages with the arm. Note that well-designed catches have a side spur or some such detail that prevents them from swiveling around out of alignment.

Fixing the Fastening and the Catch

1. Set the sprung fastening toward the top of the head stile—on the edge, so that the fixing wings are in line with the top beam—and fix it in place with the bolts.

2. With the gate in the closed position, set the catch in place against the fastening, and fix with bolts. Finally, when everything is in place, have a tryout. If you've got it right, you should be able to open the gate with little effort, and if you let go of the gate in its opening arc, it should fall back so that the sprung fastening engages with the catch.

AFTERTHOUGHTS AND OPTIONS

- Although the gate is self-closing, it's not a good idea to keep opening the gate wide and letting it fall back closed. The self-closing feature is really intended to be a fail-safe if you are worried about people leaving the gate open. It's best to close the gate manually.

- There are many types of hinges—some with wings that bolt through to the rails, some galvanized, some hand forged with fancy curlicues, and so on. If you are new to an area and are looking to install such a gate, check around the neighborhood for a local blacksmith who can make something a little bit special.

- Although kids love swinging on field gates—especially gates of the self-closing variety—there are two possible problems: The weight of the children might damage the gate, and the self-closing action might result in squashed fingers. The best advice—if your little ones really must swing on the gate—is to encourage them to swing at the harr end. This minimizes the damage by keeping the weight on the hinges.

- One of the best ways to protect a gate from the kids is to involve them in making and hanging it. If they have helped dig the hole and ram home the rubble, and maybe buried a message in the bottom of the hole, they will have a positive understanding and respond accordingly.

- Don't go to the trouble of building a beautiful oak field gate and then try to cut costs by hanging the gate on softwood posts—it's false economy.

- Bearing in mind that the tannin in oak will swiftly eat iron—resulting in black stains all over the wood and a rapid breakdown in the iron—grease the bolts before you set them in place, use galvanized bolts, or best of all, use high-quality stainless bolts.

- Some old-timers used to protect the bottoms of the posts from rot and decay by dipping them in tar or tar oil or charring them in a bonfire. The charring idea is pretty good, because the charcoal provides a thick barrier against rot and insect damage.

- A good alternative design for a wide gate that is to swing over a hard driveway is to fit a wheel to the toe of the gate.

- Some gate builders would fit a hanging post that stood about 3 feet higher than the gate and then run a cable from the top of the post to the top of the gate's head stile.

Gate Designs

DESIGN 1
If you want to create a romantic garden—lots of bowers draped in honeysuckle and sweetbrier—or if you have a fancy for pretty enclosures to separate the various areas of the garden, little rustic bower gates are the answer. They look beautiful, and because they can be built from locally grown wood—trimmings from local trees—they immediately blend in with their surroundings.

Gate type and location: Basic framed rustic bower gate; good for a traditional romantic garden setting, perhaps to separate the herb garden from the general area. Gates of this character look good in woodland gardens.
Construction and joints: Framed construction, with the primary rails mortised into the stiles and all the secondary sticks nailed to the frame. It's best if the frame is made from a long-lasting wood such as ash, so that the secondary elements can be replaced at intervals.
Skill level: A wonderfully easy gate to build. Some gate makers speed up the whole operation by boring the mortises.

DESIGN 2

The wonderful thing about traditional picket gates is that you can easily change the design by simply changing the line of the top of the pickets, the shape of the pickets, or both. Illustrated from top to bottom are round tops set in a straight line, round tops set concave, round tops set convex, and round tops set stepped.

Gate type and location: *Ledged and braced picket gate; good for a country cottage garden. In England, such gates are often associated with thatched cottages.*

Construction and joints: *A framed gate with the ledges or rails mortised and tenoned into the stiles, and the braces nailed to the rails. The pickets are nailed to the frame. Poor-quality picket gates are butt-jointed and nailed, with no mortise and tenon joints. Depending on the design, the pickets are cut with a backsaw or a coping saw. Three details decide the design of the gate: the shape of the picket (that is, the shape of the top), the spacing of the pickets, and the alignment of the tops of the pickets.*

Skill level: *A very easy gate to build, especially if you use a cutting jig to shape the picket tops.*

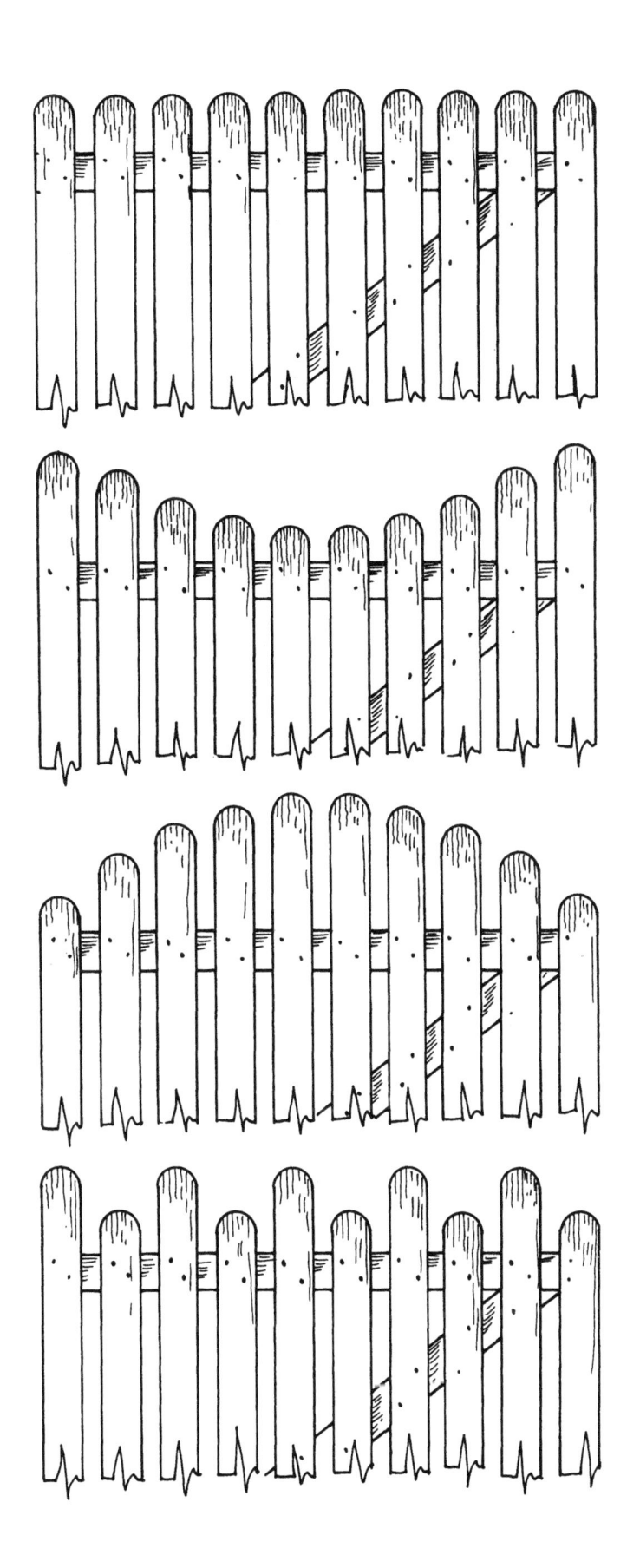

Gate Designs • 151

DESIGN 3

Although picket gates are easy to build, it's plain to see that the smallest change—in the shape of the individual tops or in the alignment of the tops—gives the gate a completely different image. These pointed tops look much more aggressive than the rounded tops.

Gate type and location: *Ledged and braced picket gate; good for traditional country cottage locations.*
Construction and joints: *A framed gate with ledges or rails; the pickets are nailed to the frame. The design has to do with the shape of the pickets, their spacing, and their height.*
Skill level: *A very easy gate to build, with the skill being in the design and arrangement of the pickets.*

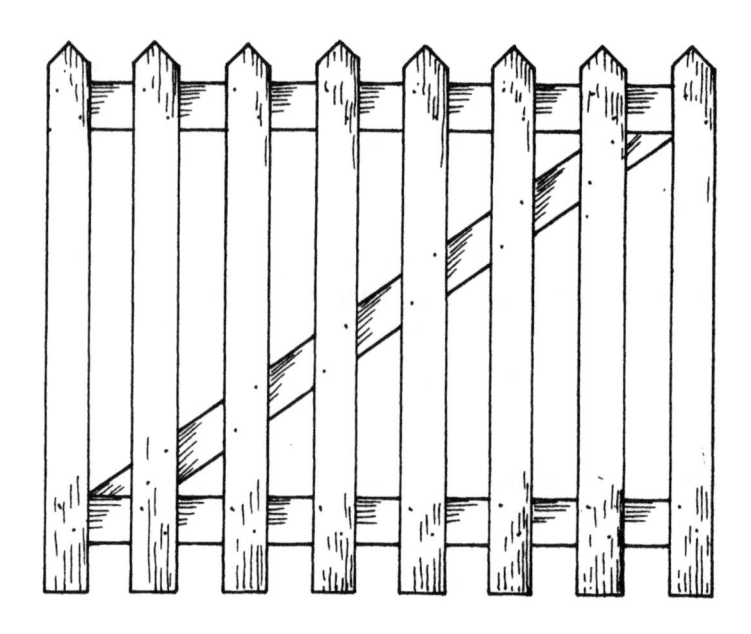

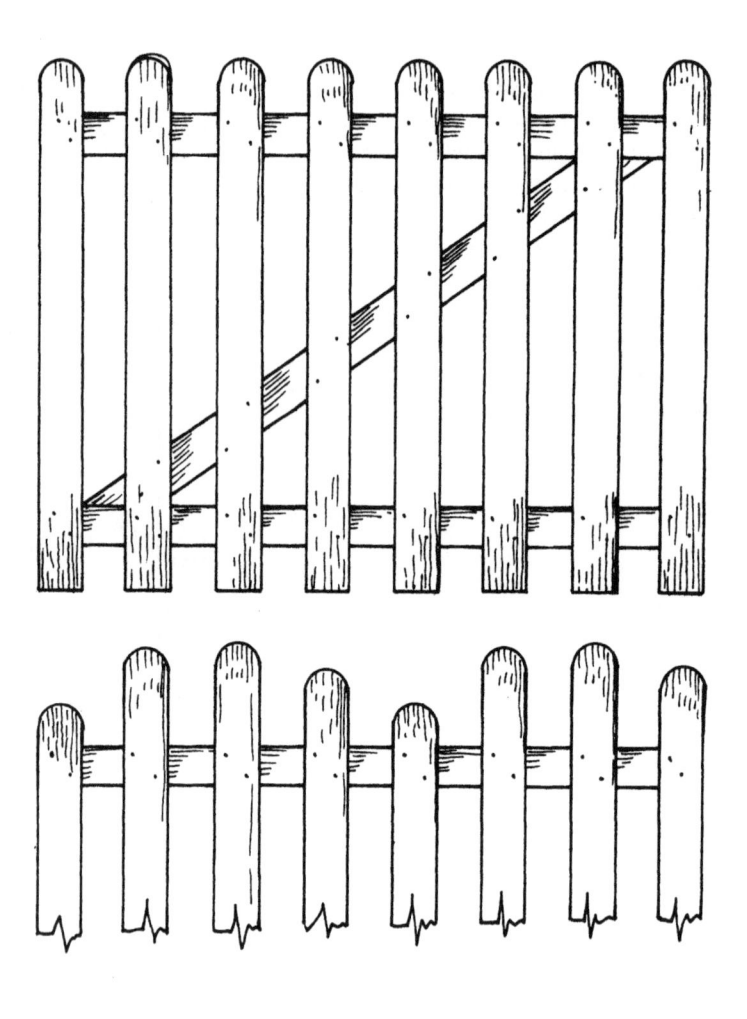

152 • *Gates*

DESIGN 4

When considering which type of picket to use, remember that straight lines are much easier to cut than curves. All the pickets illustrated can be achieved by using a backsaw to make one or more straight cuts. From top to bottom the cuts increase in number and complexity: single cut—square; single cut—mitered; double cut—pyramidal or pointed; double cut—bird's mouth or arrow notch; triple cut—spiked; six cut—spear point; nine cut—Christmas cracker.

Gate type and location: *A framed, ledged, and braced picket gate; good for a traditional cottage location.*
Construction and joints: *The rails or ledges are mortised and tenoned into the stiles, and the pickets are nailed to the frame. Note that the shape of the picket top sets the pace for the design of the gate.*
Skill level: *Although the gate is easy to build, cutting the pickets may be a chore. The art of making picket gates has to do with being able to reduce the cutting to a number of swift, straight cuts. All the designs illustrated can be achieved by making simple straight cuts.*

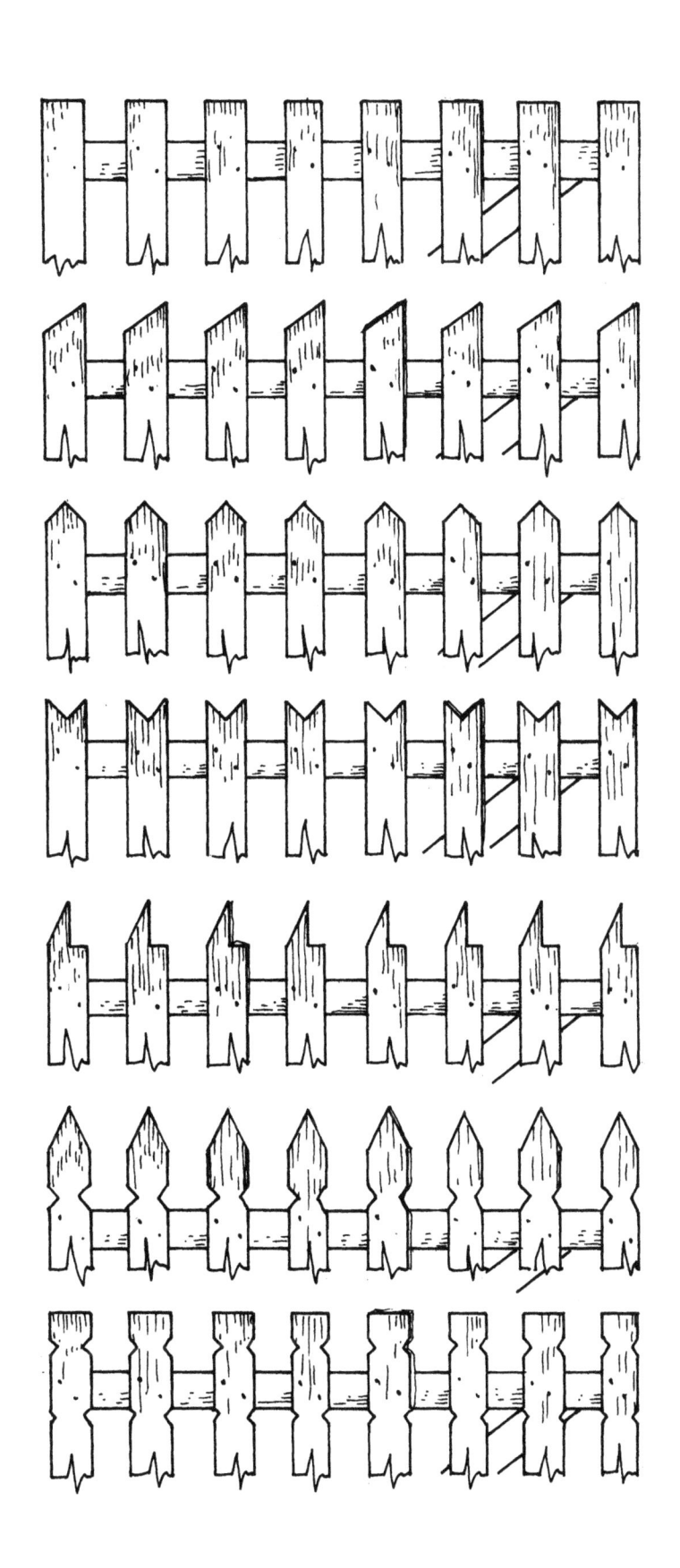

Gate Designs • 153

DESIGN 5

Although just about every American, European, and British rural area has its own traditional farm gate, the best of the bunch are the five-bar field gates found in southern England. Their beauty is the direct result of their structure, with the patterns created by the various purposes of the gate. The spacing of the rails has to do with the size of the livestock—sheep, cattle, or horses—and the number of braces and vertical slats has to do with the width and strength of the gate. For example, gates for sheep can be built with five bars, a single brace, and a single slat, whereas gates for cattle need to be heavier with a bigger top beam, two or more braces, and three or more slats.

Gate type and location: A basic framed five-bar field gate. This type of gate is traditionally used in a field or paddock situation—where you need to contain livestock—but it also makes a good driveway gate.

Construction and joints: The rails are mortised and tenoned into the stiles, the braces are notched into the stiles alongside the rails, and the whole thing is clinched together with pegs or nuts and bolts.

Skill level: This is a fairly easy gate to build. The skill required has more to do with maneuvering the long, heavy pieces of timber and hanging the gate than with actually cutting the joints.

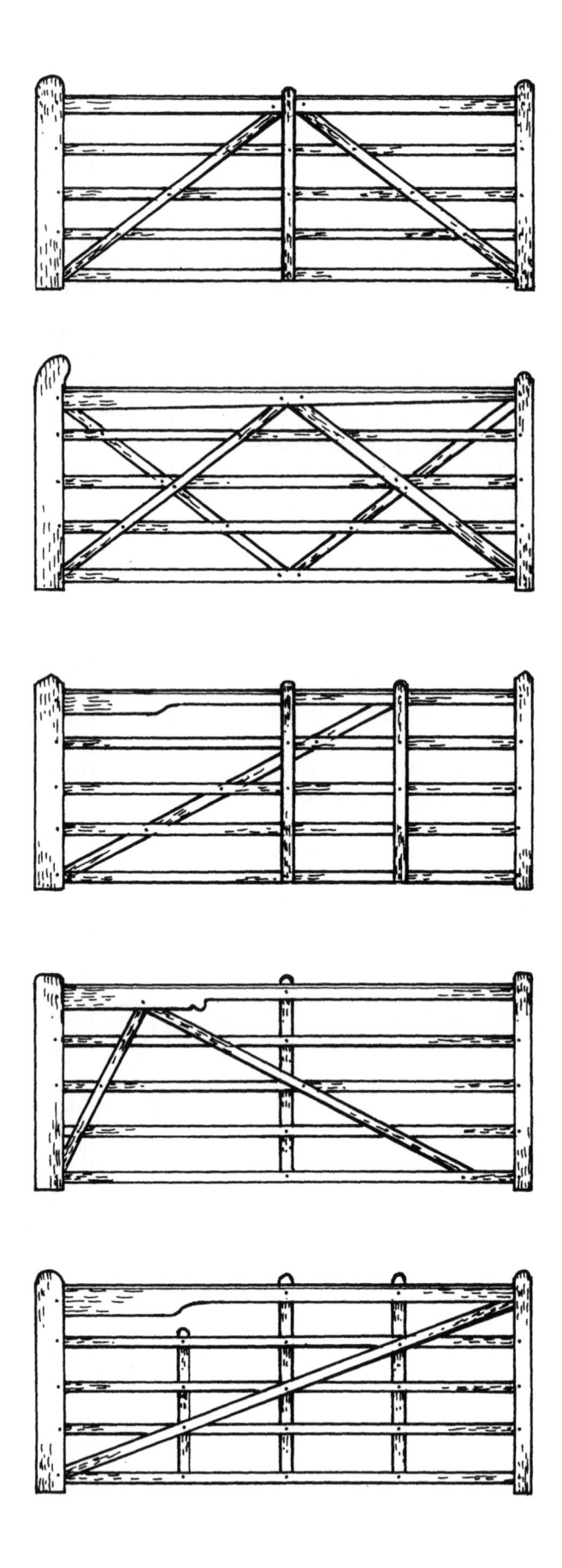

DESIGN 6

If you want to build a top-of-the-line field gate, you can't do better than a gate with a curved stile, a trussed harr, and a brace that runs from the stile down to the toe. Of course, you'll have to spend time searching for naturally shaped wood to suit the various curves and swellings of the design, but you'll wind up with a gate that will last a lifetime.

Gate type and location: *A complex curved stile field gate; good for livestock gates or carriage drives.*
Construction and joints: *A framed gate with the rails mortised and tenoned into the stiles; the braces are fitted with bare-faced tenons and laps.*
Skill level: *The skill involved in making this gate has to do with being able to select suitable timber for the harr stile. The problem is that the harr stile needs to be worked from a piece of timber that has two distinct features: a natural bend at the top, and a well-placed branch spur. The idea is that the curve at the top of the stile and the truss can both be achieved without the need to cut across the grain.*

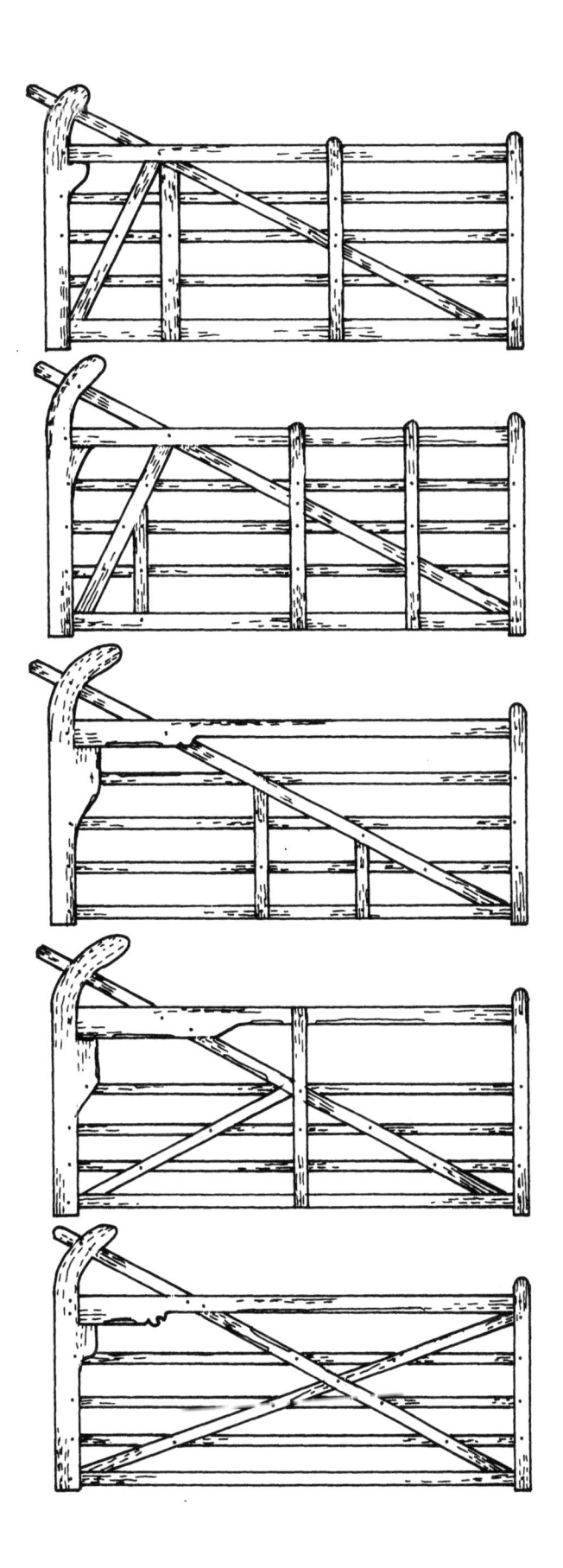

Gate Designs • 155

DESIGN 7

If you are worried about the neighbors peering into your yard or you want to keep strangers out and family and pets in, or if you simply like your privacy, perhaps a pair of high courtyard gates is the answer. These gates look great in small yards that back onto a footpath or sidewalk. The close boarding and the height mean that you can sit out in the yard and have a fair degree of privacy.

Gate type and location: Framed and battened courtyard gates; good for high-walled gardens, backyards, and courtyards.

Construction and joints: The ledges are mortised and tenoned into the stiles. Both the stiles and the ledges are rabbeted, and the panels are built from tongued and grooved battens.

Skill level: A tricky gate to build because of the need to achieve a pair of gates that are true to each other. Perhaps more than anything else, the required skill has to do with being able to cut shouldered tenons that are accurate and square.

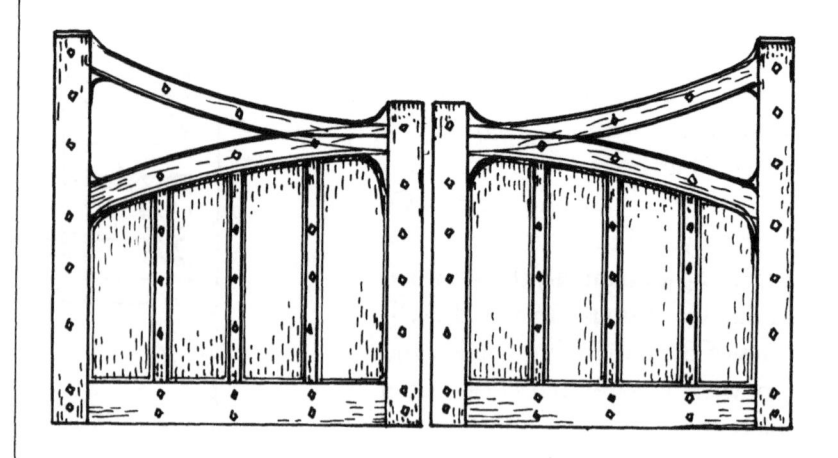

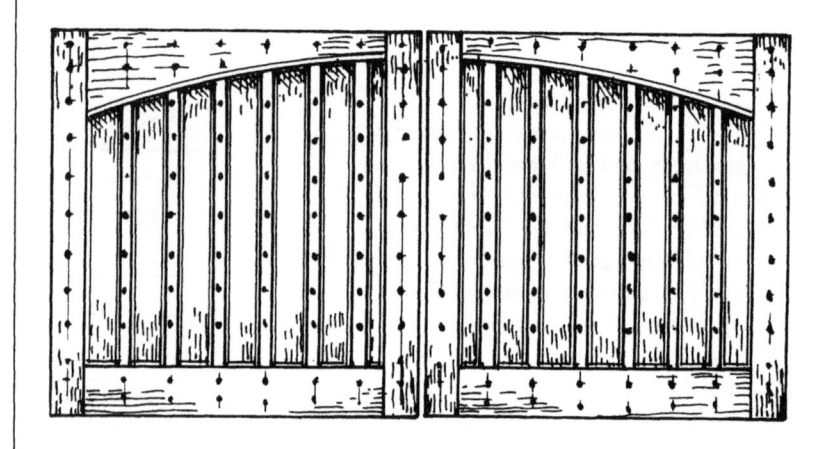

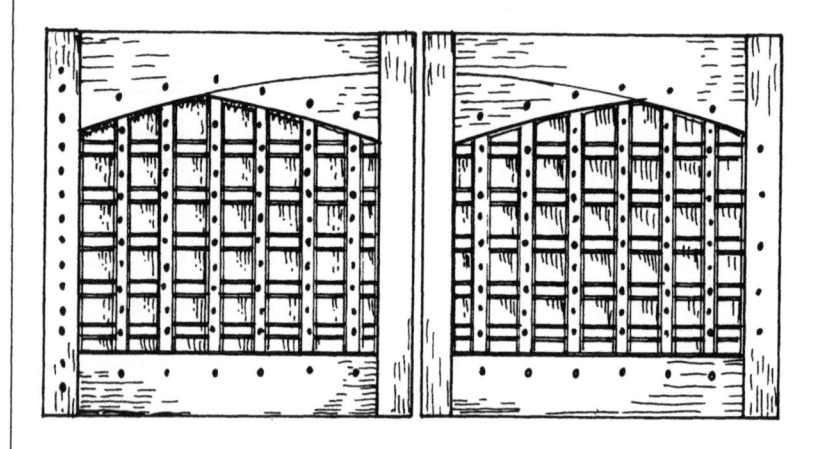

DESIGN 8

In many ways, the side or service gates illustrated are designed to fulfill the same function as courtyard gates: They keep unwanted eyes and unappreciated advances at bay. Additional pluses are the heightened design and the extra strength of the structure. Although such gates are usually situated out of the way at the back of the house and are very functional—with lots of locks, bars, and bolts on the inside of the gate—they are still highly decorative. If you need a good, solid pair of gates that looks attractive, provides security and privacy, and sends out signals of strength, permanence, and Old World traditions, these are the gates for you.

Gate type and location: A complex service gate, suitable for a house that requires a high degree of privacy and prestige.
Construction and joints: Framed and ledged; the shaped ledges and rails are mortised and tenoned into the stiles, and the muntins are grooved and stub-tenoned into the frame.
Skill level: Although the various joints are straightforward, the sum of all the joints makes for a complex structure. The required skill has to do with being able to get it all together so that everything fits.

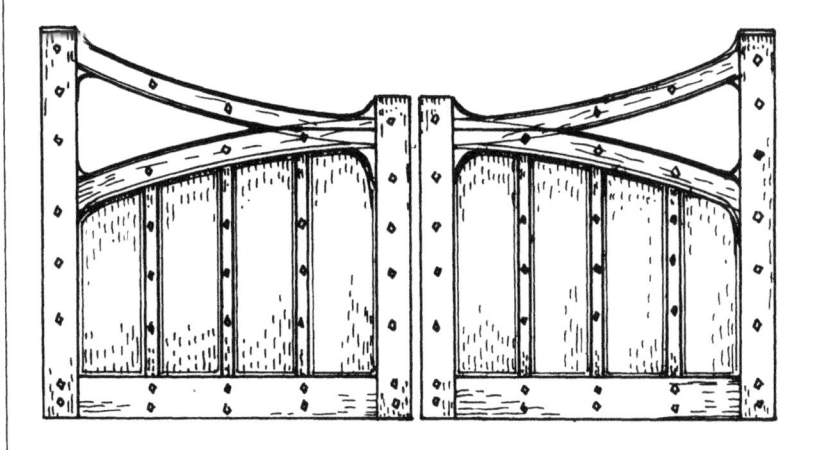

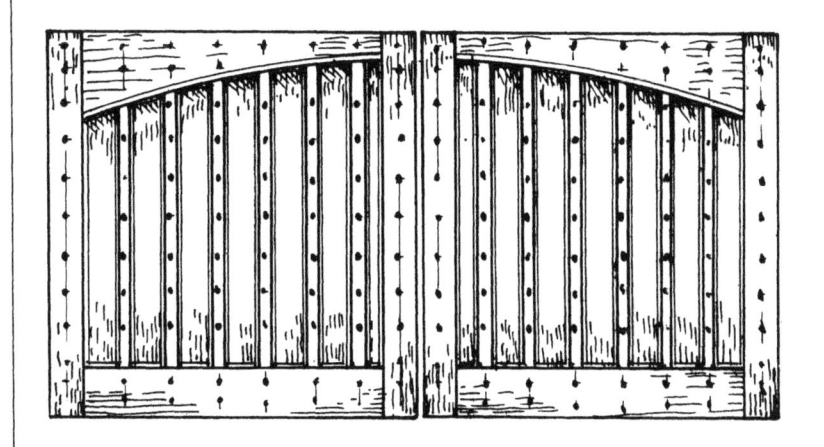

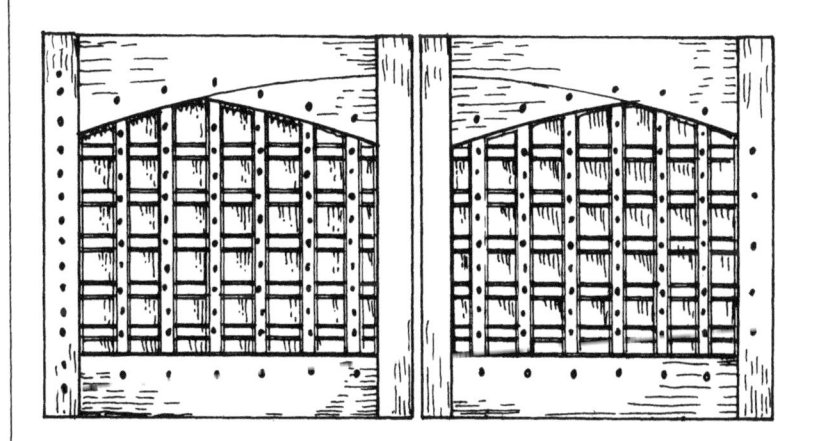

Gate Designs • 157

DESIGN 9

Small, well-built front gates of this character are perfect for small town houses—those that have minimal brick walls and gardens in front. Although these two gates were designed and built for two 1940s London houses, they are well suited for certain houses in Boston, Cape Cod, or wherever.

Gate type and location: A traditional framed, ledged, and paneled gate; good for a town house setting.
Construction and joints: Framed, ledged, and paneled; the ledges are mortised and tenoned into the stiles, the muntin is stub-tenoned into the ledges, and other details are variously turned and worked with a band saw. The top gate has square sectioned mullions between the top two rails.
Skill level: Tricky gates to build because they involve so many different techniques. For example, the bottom gate requires that you be able to turn a number of identical spindles; you also need to be familiar with joint making. You'll need both a band saw and a lathe.

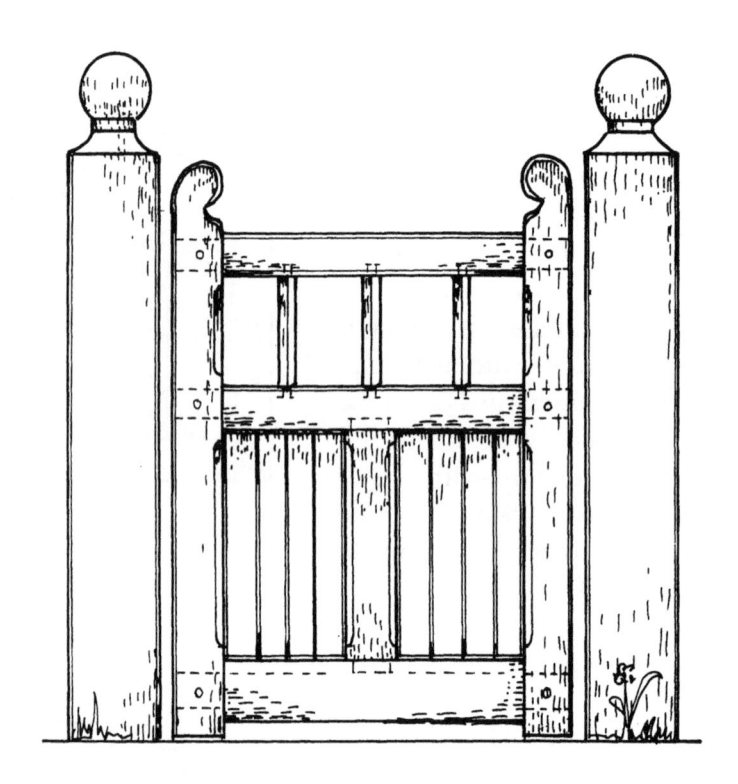

158 • *Gates*

DESIGN 10

Although lots of front gates are stylish—in the sense that they are positively designed and well built—these two gates were built to complement specific styles. The one on top was built for an Arts and Crafts house, and the one on the bottom was built for an Art Deco house. They are still conservative and understated, with all the features you would expect to see in a well-structured gate.

Gate type and location: A traditional framed gate; perfect for a small walled front garden. The gate on top—dated about 1940—was built in oak for a small London town house. The gate on the bottom was designed for an Art Deco house.
Construction and joints: Framed, ledged, and paneled, with the ledges mortised and tenoned into the stiles and the muntin stub-tenoned between the top rail and the top ledge. The frame is rabbeted to take the tongued and grooved battens.
Skill level: Although both gates are relatively easy to build, the gate with the diagonal battens is much more difficult, simply because each batten has to be mitered.

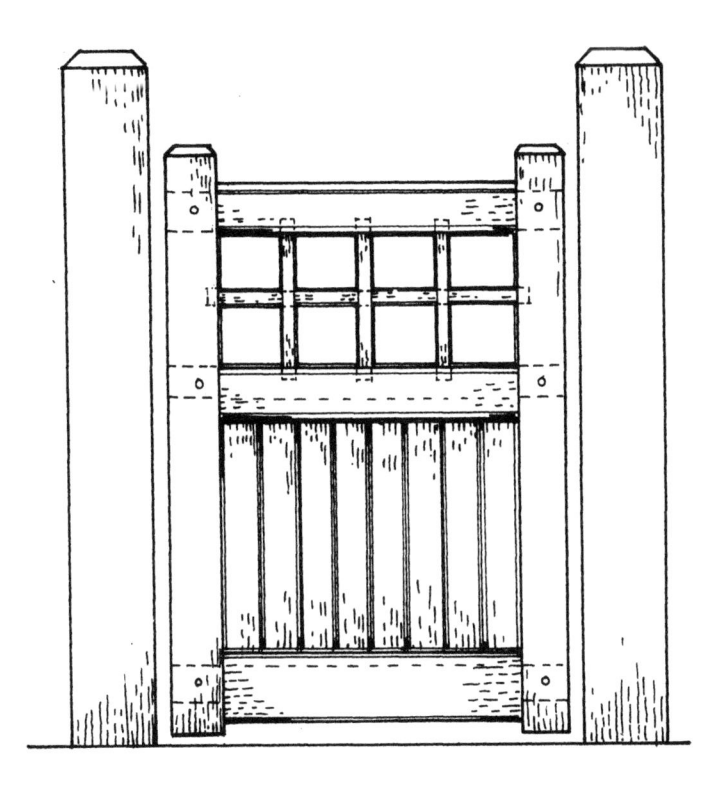

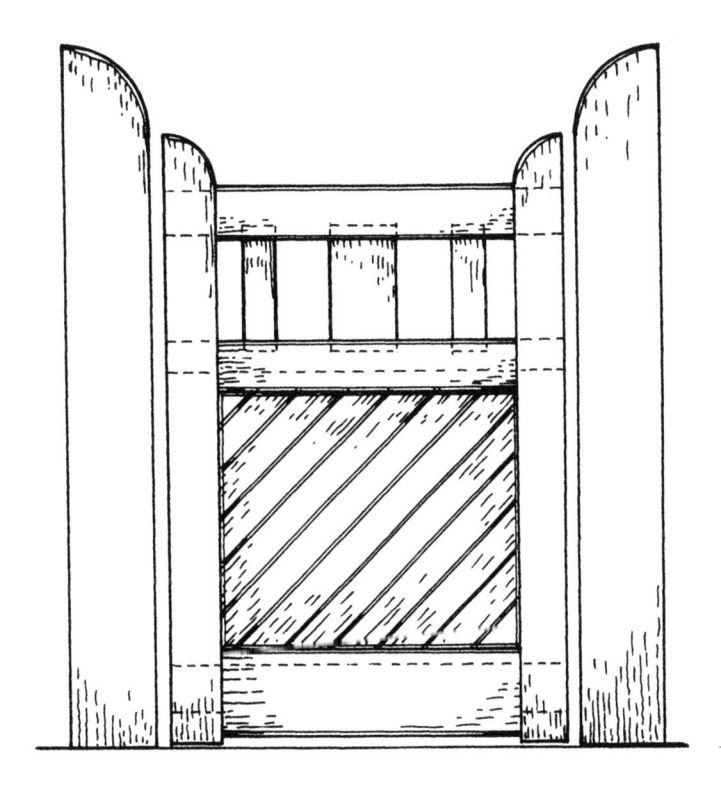

Gate Designs • 159

DESIGN 11

If you want to build a top-quality oak gate in the Arts and Crafts or Craftsman tradition, one of these Arts and Crafts gates may be for you. In many ways, they are no more than rather modest entrance gates; the flourish is provided by the braces.

Gate type and location: *Basic framed, ledged, braced, and paneled front gates, both designed specifically for Arts and Crafts houses.*
Construction and joints: *Framed, with the ledges mortised and tenoned into the stiles and the braces stub-tenoned into the frame. Made in oak throughout.*
Skill level: *These gates call for a high level of skill, because the emphasis of Arts and Crafts is a top-quality finish. The use of untreated oak means that the fitting and the joints are up-front and on show. You can't hide mistakes under filler and paint.*

DESIGN 12

Cottage gates are small, friendly, and decorative, with lots of sugary curlicues. If you have a little cottage in the country with lots of traditional details—maybe a thatched or shingle roof and roses around the door—chances are that one of these designs will fit the bill. If Hansel and Gretel's cottage had had a front gate, it would have been one of these.

Gate type and location: *Traditional framed and paneled gates, all designed for small country cottages.*

Construction and joints: *Framed and paneled gates with the ledges and rails mortised and tenoned into the stiles. Although the gates look very different from one another, it's plain to see that the difference is achieved by making relatively small changes to the basic design.*

Skill level: *Although the gates are relatively basic, you have to spend time upfront making sure that the details are right. It's best to take a good long look at your house—its style and location—and try to build related details into the design of the gate. For example, if your house is Victorian with lots of fussy detail, perhaps the gate at the bottom right is suitable.*

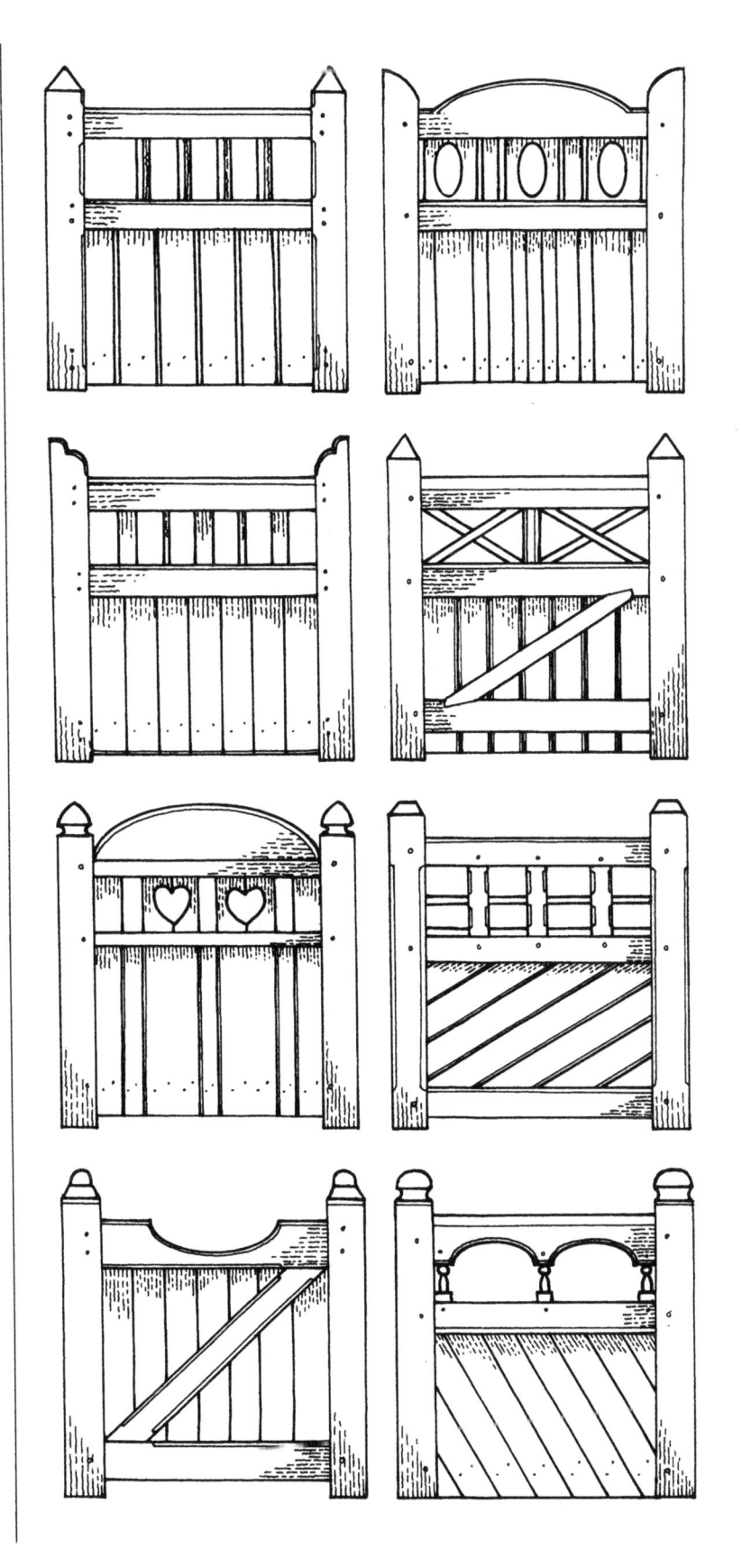

Gate Designs • 161

DESIGN 13

If you are going to all the trouble, time, and expense of building your own gate, you should set it between a pair of special gateposts. The illustrations show that the design of the gatepost is limited only by your imagination. No matter what your particular woodworking interests—band saw work, carving, turning, router work—there are a number of exciting possibilities.

Gate type and location: *Gateposts for gates in general.*
Construction: *Top, left to right:*
- *Square section post—square sawed top*
- *Square section post—sawed pyramidal top*
- *Square section post—band-sawed top with chamfers*
- *Round section post—turned half round with grooves*

Second down, left to right:
- *Square section post—band-sawed profile*
- *Round section post—turned ball finial with grooves*
- *Square section post—band-sawed bracket shape*
- *Square section post—band-sawed V-cut*

Third down, left to right:
- *Round section post—turned ball*
- *Square section post—pyramidal cap and side reeding*
- *Square section post—band-sawed quarter-circle*
- *Square section post—turned ball finial*

Bottom, left to right:
- *Square section post—turned cone and ball finial*
- *Round section post—turned ball finial and carved detail*
- *Square section post—band-sawed scroll top*
- *Square section post—carved details*

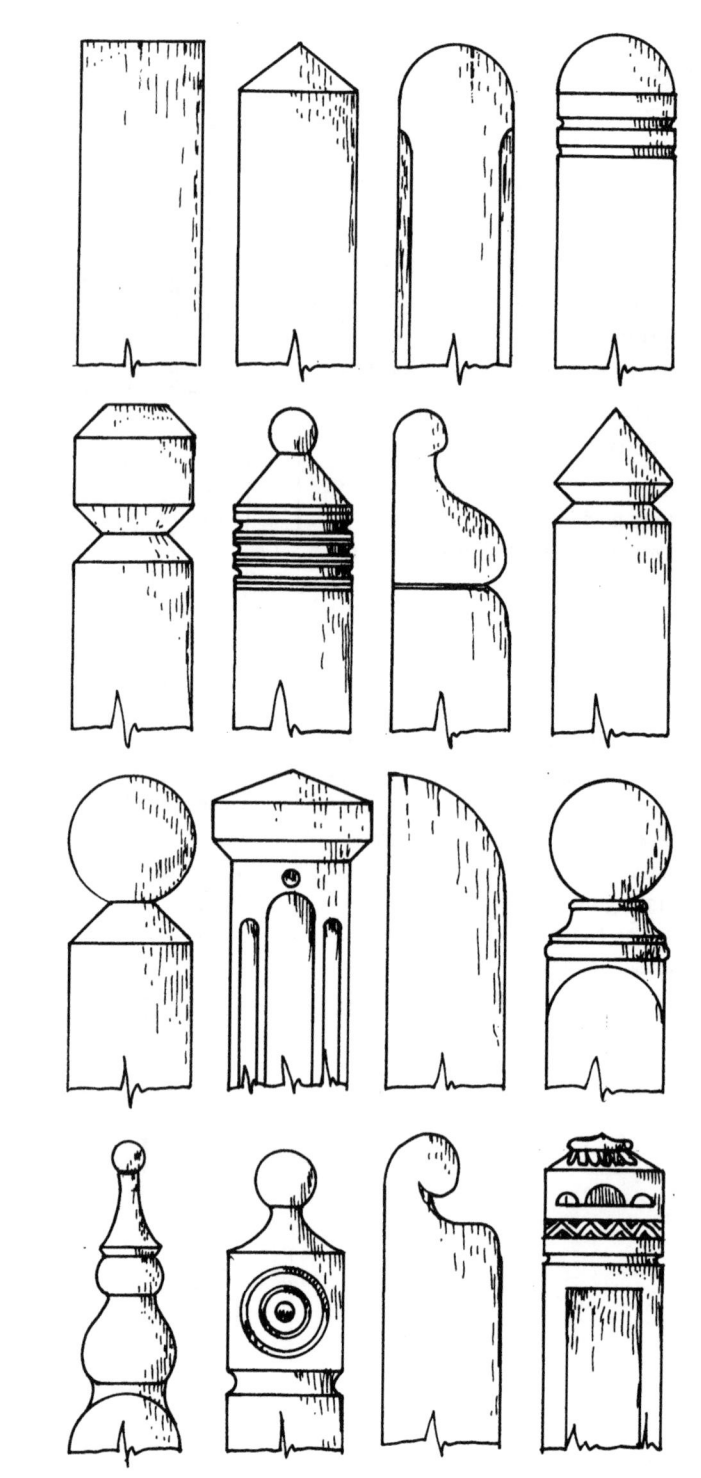

Skill level: *Although the skill level relates to which post you want to make, the turned posts are the most complex, if only because they need to be worked on a massive lathe. You can simplify the turning procedure by turning a ball (or whatever) and then fixing it to the post with a long pegged tenon.*

DESIGN 14

The question to ask is whether the gates are intended to be anything more than decorative. Certainly the gates are going to be a highly visual barrier that marks the extent of your property, but more than that, do you want to contain small children or pets? Or can you go for a more open and decorative structure? These four gates cover a whole range of needs.

Gate type and location: *Traditional framed carriage gates; best suited for a country house. The designs with the small side gate—sometimes described as a side wicket gate—are particularly suitable when there isn't room for a pair of double carriage gates and a separate side gate.*

Construction and joints: *Framed, with ledges and rails mortised and tenoned into the stile, and with pickets and panels.*

Skill level: *In essence, the four designs are all basic framed gates. The third gate down—the one with diagonal members—is a difficult challenge. Not only do all the diagonal members need to be stub-tenoned into the frame, but they also have to be parallel to one another and well spaced within the frame.*

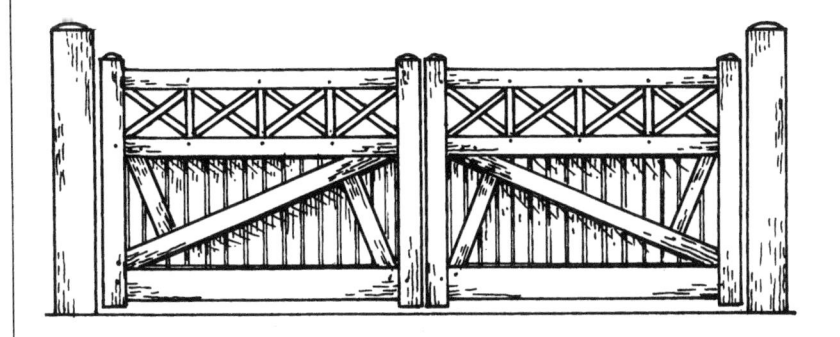

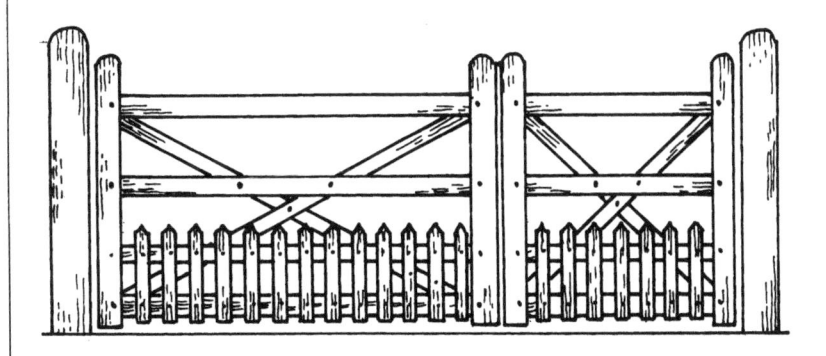

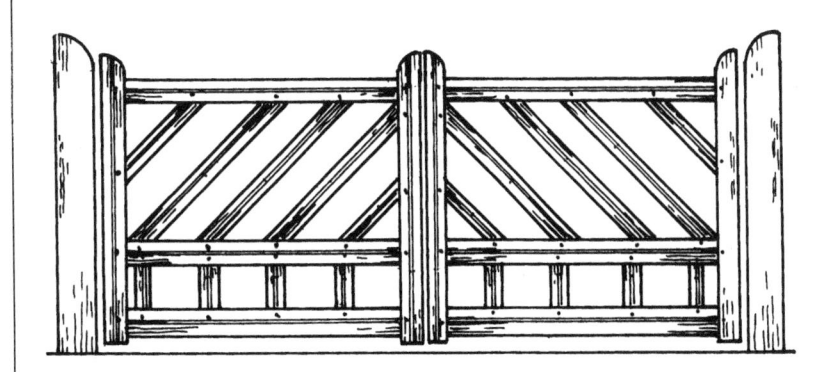

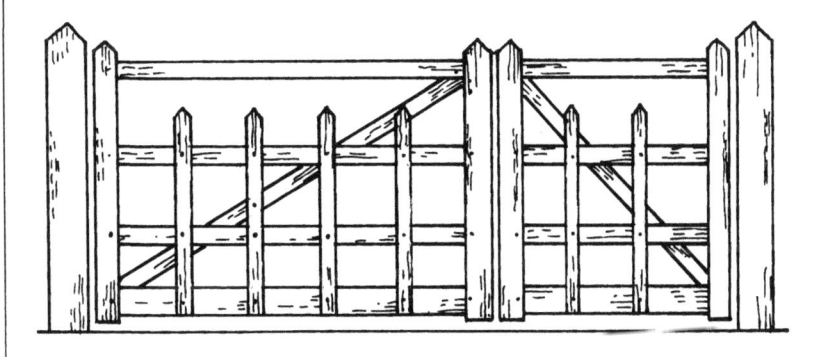

Gate Designs • 163

DESIGN 15

These carriage gates have been designed specifically to keep kids, pets, and toys contained. Tiny tots can't wriggle through or get their heads stuck between the bars, small dogs can't see through or jump over, and the space between the bottom of the gate and the surface of the drive is such that toys can't roll under.

Gate type and location: *Framed, ledged, paneled, and variously braced drive gates designed for formal country and town houses.*

Construction and joints: *Framed, with the ledges mortised and tenoned into the stiles. Some designs have braces, and others have muntins, but all the gates have open mullions at the top.*

Skill level: *In many ways, gates like this are the most difficult to build, if only because there are so many individual components and so many techniques. For example, with the second gate down, it's not that turning the spindles is so difficult, or that cutting the fancy tops on the stiles is too much of a problem, or even that turning the post balls is tricky; but getting all these techniques together in one project requires a great deal of organizing. The best way to proceed is to study the designs, do a bit of research by looking at other people's gates, and then make detailed working drawings.*

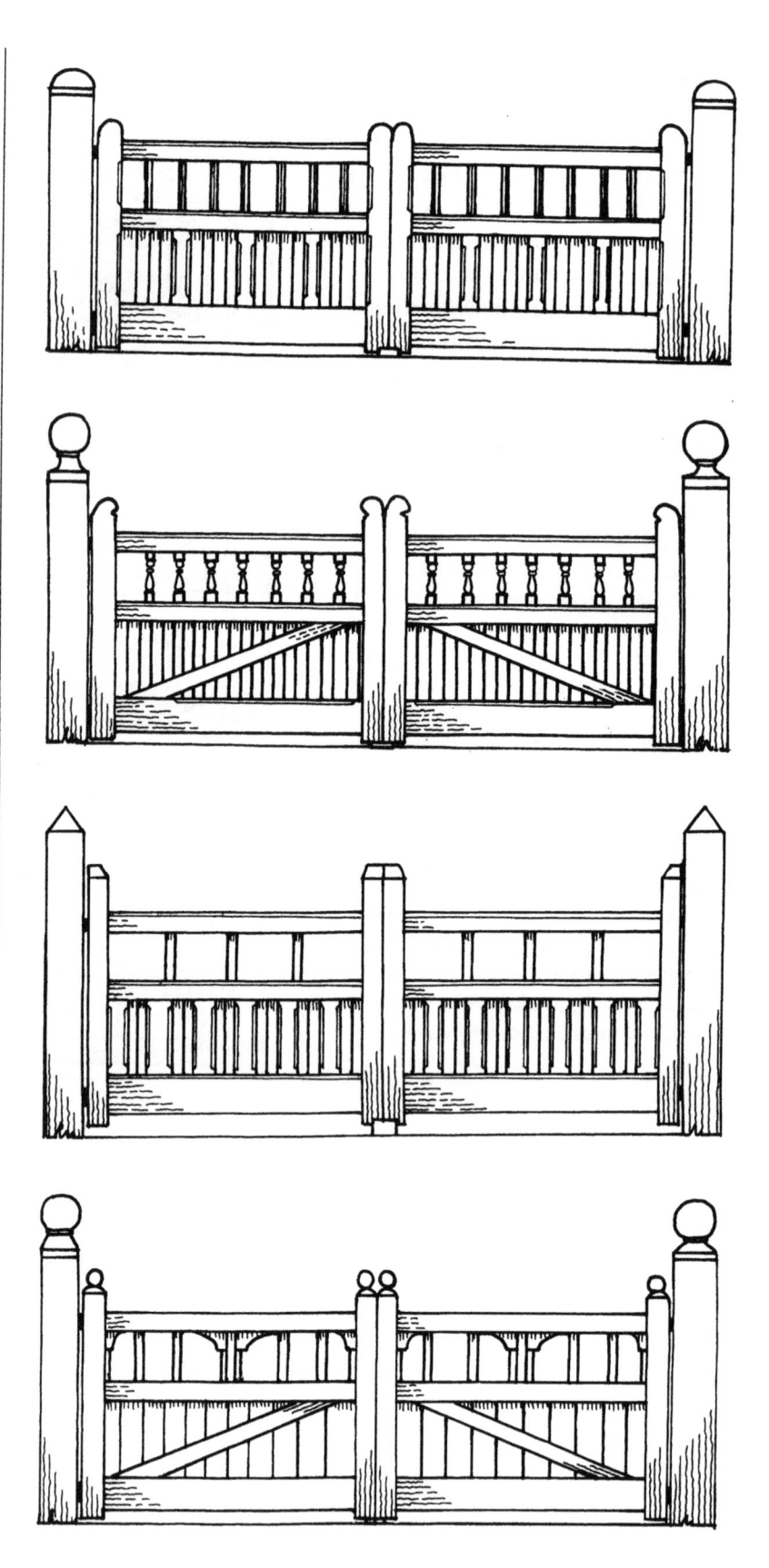

METRIC CONVERSIONS

INCHES TO MILLIMETERS

in.	mm	in.	mm
1	25.4	51	1295.4
2	50.8	52	1320.8
3	76.2	53	1346.2
4	101.6	54	1371.6
5	127.0	55	1397.0
6	152.4	56	1422.4
7	177.8	57	1447.8
8	203.2	58	1473.2
9	228.6	59	1498.6
10	254.0	60	1524.0
11	279.4	61	1549.4
12	304.8	62	1574.8
13	330.2	63	1600.2
14	355.6	64	1625.6
15	381.0	65	1651.0
16	406.4	66	1676.4
17	431.8	67	1701.8
18	457.2	68	1727.2
19	482.6	69	1752.6
20	508.0	70	1778.0
21	533.4	71	1803.4
22	558.8	72	1828.8
23	584.2	73	1854.2
24	609.6	74	1879.6
25	635.0	75	1905.0
26	660.4	76	1930.4
27	685.8	77	1955.8
28	711.2	78	1981.2
29	736.6	79	2006.6
30	762.0	80	2032.0
31	787.4	81	2057.4
32	812.8	82	2082.8
33	838.2	83	2108.2
34	863.6	84	2133.6
35	889.0	85	2159.0
36	914.4	86	2184.4
37	939.8	87	2209.8
38	965.2	88	2235.2
39	990.6	89	2260.6
40	1016.0	90	2286.0
41	1041.4	91	2311.4
42	1066.8	92	2336.8
43	1092.2	93	2362.2
44	1117.6	94	2387.6
45	1143.0	95	2413.0
46	1168.4	96	2438.4
47	1193.8	97	2463.8
48	1219.2	98	2489.2
49	1244.6	99	2514.6
50	1270.0	100	2540.0

The above table is exact on the basis: 1 in. = 25.4

U.S. TO METRIC

1 inch = 2.540 centimeters
1 foot = .305 meter
1 yard = .914 meter
1 mile = 1.609 kilometers

METRIC TO U.S.

1 millimeter = .039 inch
1 centimeter = .394 inch
1 meter = 3.281 feet or 1.094 yards

INCH-METRIC EQUIVALENTS

Fraction	Decimal Equivalent Customary (in.)	Metric (mm)	Fraction	Decimal Equivalent Customary (in.)	Metric (mm)
1/64	.015	0.3969	33/64	.515	13.0969
1/32	.031	0.7938	17/32	.531	13.4938
3/64	.046	1.1906	35/64	.546	13.8906
1/16	.062	1.5875	9/16	.562	14.2875
5/64	.078	1.9844	37/64	.578	14.6844
3/32	.093	2.3813	19/32	.593	15.0813
7/64	.109	2.7781	39/64	.609	15.4781
1/8	.125	3.1750	5/8	.625	15.8750
9/64	.140	3.5719	41/64	.640	16.2719
5/32	.156	3.9688	21/32	.656	16.6688
11/64	.171	4.3656	43/64	.671	17.0656
3/16	.187	4.7625	11/16	.687	17.4625
13/64	.203	5.1594	45/64	.703	17.8594
7/32	.218	5.5563	23/32	.718	18.2563
15/64	.234	5.9531	47/64	.734	18.6531
1/4	.250	6.3500	3/4	.750	19.0500
17/64	.265	6.7469	49/64	.765	19.4469
9/32	.281	7.1438	25/32	.781	19.8438
19/64	.296	7.5406	51/64	.796	20.2406
5/16	.312	7.9375	13/16	.812	20.6375
21/64	.328	8.3384	53/64	.828	21.0344
11/32	.343	8.7313	27/32	.843	21.4313
23/64	.359	9.1281	55/64	.859	21.8281
3/8	.375	9.5250	7/8	.875	22.2250
25/64	.390	9.9219	57/64	.890	22.6219
13/32	.406	10.3188	29/32	.906	23.0188
27/64	.421	10.7156	59/64	.921	23.4156
7/16	.437	11.1125	15/16	.937	23.8125
29/64	.453	11.5094	61/64	.953	24.2094
15/32	.468	11.9063	31/32	.968	24.6063
31/64	.484	12.3031	63/64	.984	25.0031
1/2	.500	12.7000	1	1.000	25.4000

Index

Numerals in italics indicate illustrations

abacus, 7
apron rail, *11*
architrave, 6, *13*
arris, *10*
Art Deco door, 99, *99*
Arts and Crafts style front gates, 160, *160*

bare-faced tenons, 16
basic doors, 3
batten, 5
batten doors
  anatomy of, *5*
  see also framed, ledged, and braced doors; ledged and braced doors
batten and ledged doors, traditional, 3, 82, *82*
bead, 9
boarded gate, 117, *117*
bolection molding, 7, *13*
bower gates, 118, *118*
  rustic, 150, *150*
brace, 8
butts, charred, 112

carriage gates, 118, *118*
  with combined wicket gate, 119, *119*
  Edwardian braced and boarded, 120, *120*
  framed, ledged, paneled, and braced, 164, *164*
  traditional framed, 163, *163*
casings. See door linings
cavetto, *13*
chamfer, *12*
Church of All Saints, High Roding, Essex, England, 77
church doors, framed, ledged, and braced, 28
Church of St. Michael, Fobbing, Essex, England, 79
cornice, 7, 9
cottage doors, 3

cottage gates
  sawed finial Victorian, 126, *126*
  Swiss, 129, *129*
  traditional framed and paneled, 161, *161*
courtyard gates, *104*, 119, *119*
  framed and battened, 156, *156*
cross-tongue, 16
curved stile narrow entrance gates, 120, *120*
cyma curve, *11*

dado, 17
dentils, 7
diminished stile doors, 39, *39*
diminished stile mortise and tenon, 17
door designs
  Art Deco three-plank, 99, *99*
  basic framed and paneled, 84, *84*, 86, *86*, 91, *91*
  basic slab plank, 72–74, *72–74*, 78–79, *78–79*
  Eastlake style framed and paneled, 93, *93*
  framed and braced garage, 101, *101*
  framed and glazed, 100, *100*
  framed and ledged sash, 88, *88*
  framed and paneled, 82, *82*
  framed and paneled with built-up curved heads, 94, *94*
  framed and paneled sash, 92, *92*
  framed and paneled with a transom window, 95, *95*
  late-eighteenth-century framed and paneled, 87, *87*
  neo-Georgian basic framed and paneled, 98, *98*
  neo-Jacobean framed and paneled, 100, *100*
  1920's framed and paneled, 97, *97*
  1940s flush, 100, *100*
  slab plank, 75–77, *75–77*
  slab plank with lancet head, 77, *77*
  traditional batten and ledged, 82, *82*
  traditional framed and paneled, 83, *83*, 85, *85*, 89, *89*, 96, *96*
  traditional framed and paneled with decorative glass, 90, *90*
  traditional plank and ledge, 80, *80*
door fixtures, 5, 60–63, 68–70, *69–71*
  traditional arrangement for, 60
door-frame dovetail, 17
door frames, 54–55, *55–56*
  anatomy of, *9*
door joints, 5, 16–23
  bare-faced tenons, 16
  cross-tongue, 16
  dado, 17
  diminished stile mortise and tenon, 17
  door-frame dovetail, 17
  double-paired tenons, 18
  folding wedges, 18
  foxtail wedging, 18
  framed, 19
  groove, 20
  haunched tenon, 20
  miter, 20
  oblique mortise and tenon, 21
  paired tenons, 21
  pinned tenon, 21
  rabbet, 22
  scribed, 22
  stub tenon, 23
  tongued and grooved panel, 23
door linings, anatomy of, *10*
doors, vii–viii, 1
  batten, 5
  batten and ledged, 3
  battened ledged and braced, 3
  design and construction, 1–4
  diminished stile, 39, *39*
  double-margin, 37, *37*
  exterior, 39
  four-panel, 45–53
  framed and cross-battened, 31, *31*

framed and cross-braced, 30, *30*
framed and ledged, 26, *26*
framed, ledged, and braced, 27–29, *27–29*
framed and paneled, 3, 6–7
framed and paneled, four-panel, 35, *35*
framed and paneled, simple, 32–34, *32–34*
framed and paneled, six-panel, 36, *36*
framed and paneled, with carved surround, *2*
furnishing, 68–71
hanging, 63–71
interior, 39
ledged, 24, *24–25*, 40–44
ledged and braced, *3*, 25, *25*
sash, *3*, 38, *38*
Victorian period, 85, *85*
see also door designs; doors, anatomy of
doors, anatomy of, 5–15
Dore Abbey, Hereford, England, 78
double entrance doors, 37–38, 90, *90*
double-margin doors, 37, *37*
double-paired tenons, *18*

Eastlake style doors, 93, *93*
Edwardian braced and boarded carriage gates, 120, *120*
English country park gates, 121, *121*
escutcheon. *See* scutcheon
exterior doors, 39

fan light, *7*
farm gates. *See* field gates
field gates
  complex curved stile, 155, *155*
  five-bar, *121*, 121–22, 133–39, 154, *154*
field gates, hanging, 145
  afterthoughts and options for, 149
  cutting list and fixtures for, 146
  fixing the hinges, 147–49, *147–49*
  instructions for, 146–47, *146–47*
  tools for, 145
fielded panel, *7*, *14*
fingerplate, *6*
fittings, door. *See* door fixtures
five-bar field gates, *121*, 121–22, 133, *134*, 154, *154*
  afterthoughts and options for, 139
  cutting list and fixtures for, 133
  instructions for, 134–39, *134–39*
  tools for, 133
fixtures, door. *See* door fixtures
flat panel, *6*
flush door, 1940s, 100, *100*
flush panel, *11*
folding wedges, *18*
four-panel doors, 45, 46
  afterthoughts and options for, 52–53
  cutting list and fixtures for, 46

instructions for, 47–53, *48–52*
  tools for, 45–46
foxtail wedging, *18*
framed and battened courtyard gates, 156, *156*
framed and braced doors, garage, 101, *101*
framed carriage gates, traditional, 163, *163*
framed and cross-battened doors, 31, *31*
framed and cross-braced doors, 30, *30*
framed door joints, *19*
framed gate, traditional, 159, *159*
framed and glazed doors, *3*, 100, *100*
framed and ledged doors, 26, *26*
  Victorian Gothic sash, 88, *88*
framed, ledged, and braced doors, *5*, 27–29, *27–29*
framed, ledged, and paneled gates, traditional, 158, *158*
framed, ledged, paneled, and braced drive gates, 164, *164*
framed oak garden gate, 140, *141*
  afterthoughts and options for, 144
  cutting list and fixtures for, 140
  instructions for, 141–44, *141–44*
  tools for, 140
framed and paneled cottage gates, traditional, 161, *161*
framed and paneled doors, *3*, 82, *82*
  anatomy of, 6–7
  basic, 84, *84*, 86, *86*, 91, *91*
  with built-up curved heads, 94, *94*
  Eastlake style, 93, *93*
  four-panel, 35, *35*
  late-eighteenth-century, 87, *87*
  neo-Georgian basic, 98, *98*
  neo-Jacobean, 100, *100*
  1920's style, 97, *97*
  sash, 92, *92*
  simple, 32–34, *32–34*
  six-panel, 36, *36*
  traditional, 83, *83*, 85, *85*, 89, *89*, 96, *96*
  traditional with decorative glass, 90, *90*
  with a transom window, 95, *95*
frieze, *6*
frieze panels, *7*
frieze rail, *7*
front gates, basic framed, ledged, braced, and paneled, 160, *160*
furnishings, door. *See* door fixtures

garage doors, 101, *101*
garden gates, framed oak, 140, *141*
  afterthoughts and options for, 144
  cutting list and fixtures for, 140
  instructions for, 141–44, *141–44*
  tools for, 140
gate designs
  basic framed, ledged, braced, and paneled front, 160, *160*

complex curved stile field, 155, *155*
  complex service, 157, *157*
  five-bar field, 154, *154*
  framed and battened courtyard, 156, *156*
  framed, ledged, and braced picket, 153, *153*
  framed, ledged, paneled, and braced drive, 164, *164*
  gateposts, 162, *162*
  ledged and braced picket, 151–52, *151–52*
  rustic bower, 150, *150*
  traditional framed, 159, *159*
  traditional framed carriage, 163, *163*
  traditional framed, ledged, and paneled, 158, *158*
  traditional framed and paneled cottage, 161, *161*
gate-making joints
  groove, *114*
  haunched and pegged tenon, *114*
  lap joint, *114*
  muntin and brace tenon, *114*
  pegged blind mortise, *115*
  rabbet, *115*
  slat dovetail, *116*
  tenoned and wedged jowl, *116*
  tongued and grooved battens, *116*
gateposts, 162, *162*
gates, vii–viii, 103
  boarded, 117, *117*
  bower, 118, *118*
  carriage, 118, *118*
  carriage, with combined wicket gate, 119, *119*
  courtyard, 104, 119, *119*
  curved stile narrow entrance, 120, *120*
  design and construction, 103–6
  Edwardian braced and boarded carriage, 120, *120*
  English country park, 121, *121*
  five-bar field, *105*, *121*, 121–22, 133–39, *134*
  framed oak garden, 140–44
  front, *105*
  hunting, 122, *122*
  kissing, 123, *123*
  lych, *104*, 123, *123*
  new garden estate, 124, *124*
  pale diamond lattice, 124, *124*
  palisade picket, 125, *125*
  riven, 125, *125*
  rustic, 126, *126*
  sawed finial Victorian cottage, 126, *126*
  self-closing, 127, *127*
  service, 127, *127*
  stile, 128, *128*
  Sussex field, 128, *128*, 133–39, *134*
  Swiss cottage, 129, *129*
  trussed harr field, 129, *129*

Index • 167

gates, *continued*
   turned balustrade, 130, *130*
   turnstile, 130, *130*
   Victorian Gothic, 131, *131*
   wattle-hurdle, 131, *131*
   weatherboarded, 132, *132*
   wicket, 132, *132*
   see also gate designs; gates, anatomy of
gates, anatomy of, 107–13
groove, *20*
gunstock stile doors. *See* diminished stile doors

handles, 62–63
hanging doors, 63–64
   afterthoughts and options for, 71
   instructions for, 64–68, *64–68*
   materials and fixtures for, 64
   tools for, 64
hanging stile, *7*
haunch and haunching, *20*
haunched tenon, *20*
head, *9*
hinges, 55–60
   cross-garnet (tee), 58, *58–59*
   H or parliament, 59, *59*
   knob-pin, 57, *57*
   L, 59, *59*
   lift-off butt, 56, *57*
   ram's horn, 59, *59–60*
   removing door from, *57*
   rising butt, 57, *58*
   straight butt, 56, *57*
   strap, 5, 58, *58*
horns, *9*
hunting gate, 122, *122*
hurdle gate. *See* riven gate

interior doors, 39

Jacobean door, 96, *96*
jamb, *10*
jamb lining, *10*

Kempley Church, Gloucester, England, 74
kissing gate, 123, *123*
knobs, 62–63
   installing, 70–71, *71*
knockers, 63
   cast brass, 63

latches, thumb (Norfolk, Suffolk), *5*, 60, *61*
ledged and braced doors, *3*, *5*, 25, *25*
ledged doors, 24, *24–25*, 40
   afterthoughts and options for, 44
   cutting list and fixtures for, 40–41
   instructions for, 41–43, *41–44*
   tools for, 40
ledges, *8*

lock rail, *6*
locks
   cylinder, 63
   mortise, 62, *62*, 68–70, *69–71*
   rim, 60–61, *61*
   stock, 62, *62*
lych-gates, 104, 123, *123*

mail slots, 63
match boards, *5*
meeting stiles, *7*
metric conversions tables, 165
middle hall doors, 92, *92*
miter, *20*
moldings, 13–14

neo-Georgian style doors, 98, *98*
neo-Jacobean door, 100, *100*
new garden estate gate, 124, *124*
Norfolk latches. *See* latches, thumb

oblique mortise and tenon, *21*

paired tenons, *21*
pale diamond lattice gate, 124, *124*
palisade picket gate, 125, *125*
paneled doors, basic, *3*
panels
   fielded, *7*
   flat, *6*
   frieze, *7*
pediments, *6*
   designs, 81, *81*
picket gates
   framed, ledged, and braced, 153, *153*
   ledged and braced, 151–52, *151–52*
   palisade, 125, *125*
pilasters, *7*
pinned tenon, *21*
plank and ledge doors, traditional, 80, *80*
planted molding, *14*
portico, *7*
property plans, 106

rabbet, *22*
racking, *8*
rails
   frieze, *7*
   lock, *6*
   top, *6*
rear entrance door, 100, *100*
riven gate, 125, *125*
rod, *15*
rustic gate, 126, *126*

St. Alban's Cathedral, England, 75
St. Botolph's Church, Hadstock, Essex, England, 72
St. George's Inn, Southwark, London, England, 80

St. Laurence and All Saints Church, Eastwood, Essex, England, 76
sash doors, 38, *38*
   basic, *3*
   double, 91, *91*
   framed and paneled, 92, *92*
   Victorian Gothic, 88, *88*
sawed finial Victorian cottage gate, 126, *126*
scribed joint, *22*
scutcheon, *6*
self-closing gate, 127, *127*
service gates, 127, *127*
   complex, 157, *157*
shutting stile, *6*
skirting, *6*, *14*
slab plank doors, 75–77, *75–77*
   basic, 72–74, *72–74*, 78–79, *78–79*
   with lancet head, 77, *77*
stable doors, 82
   framed ledged, and braced, 29
stile doors, diminished, 39, *39*
stile field gates, complex curved, 155, *155*
stile gate, 128, *128*
stiles
   hanging, *7*
   meeting, *7*
   shutting, *6*
stops, *10*
strap hinges, *5*
stub tenon, *23*
Sussex field gates, 128, *128*, 133, *134*
   afterthoughts and options for, 139
   cutting list and fixtures for, 133
   instructions for, 134–39, *134–39*
   tools for, 133
swing wicket gate. *See* kissing gate
Swiss cottage gate, 129, *129*

three-plank door, Art Deco, 99, *99*
tongued and grooved panel joint, *23*
top rail, *6*
tower bolt, 60, *61*
transom, *7*
trussed harr field gate, 129, *129*
turned balustrade gate, 130, *130*
turnstile gate, 130, *130*, *see also* kissing gate

vestibule doors, 94, *94*
   inner, 100, *100*
Victorian Gothic gate, 131, *131*
Victorian Gothic sash door, 88, *88*
Victorian period doors, 85, *85*

wattle-hurdle gate, 131, *131*
weatherboarded gate, 132, *132*
wicket gate, 132, *132*
wooden plugs, *15*